2014 WORLD INTERIOR DESIGN

OFFICE SPACES

国际室内设计年鉴 办公

天津华厦建筑设计有限公司　策划
[西] 卡尔斯·布鲁托　编
凤凰空间　译

江苏凤凰科学技术出版社

CONTENTS 目录

004 Anonimo Offices
010 EMC Office
016 UL Guangzhou Office
022 CONI ROVESCI
026 Industrial Association
032 The Long Barn Studio
038 BCP Offices
042 Invescon Offices
050 Opman Offices
054 Gonzalo Mardones Viviani Arquitectos Studio
058 Office Herengracht, Amsterdam, the Netherlands
062 Luminare
066 Skin Peep
072 Office Plajer & Franz Studio
078 Leo Burnett Office
084 RMS California
088 UXUS Office
092 Brunete's Office
096 Design Army Headquarters
100 Bogart Lingerie
106 RI Offices
114 Acbc Office
122 Office Exactusensu
126 Kantar World Panel
132 Offices In Torre Murano

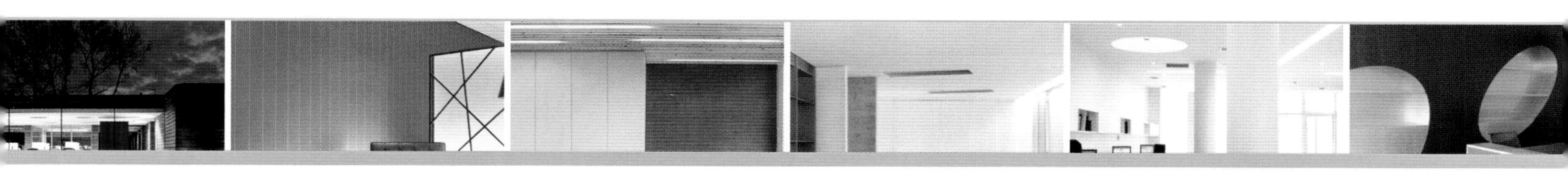

138 Prodigy MSN

144 Government Officebuilding Utrecht

152 Glem Offices

158 BAJAJ CORP OFFICE

164 GE Energy Financial Services Headquarters

172 VictorinoxSwiss Army Brands, Inc

178 Toto ish 2011

186 Cogeco Headquarters

190 Uniflair

196 Horus Capital Offices

202 Kao Corporation Head Office

208 Fox Latinamerican Channel Offices

214 Design Offices

218 1-10design Kyoto Office

224 KAYAC Ebisu Office

228 SO Architecture Office

234 Ericsson Office(Innovation Room)

238 Headquarters Office of Beijing Wintop

244 Dell Tokyo Office and Show Room

250 M&C SAATCHI Advertising Agency

258 Office Ogilvy

264 Office Besturenraad BKO

270 Agency PUBLICMOTOR

276 TOYOTA GAZOO.COM VIP ROOM

282 Paga Todo

Anonimo Offices

01

Design Company: DPGarquitectos

Project Location: Lomas de Chapultepec, Mexico City

Area: 712 m^2

Photographer: Héctor Armando Herrera

The design for the firm Anónimo responded to the need to adapt an old Mexico City home to the complexity and programmatic dynamism of a youthful advertising agency. A project priority was to respect the inherent integrity and characteristic features of the existing structure without sacrificing the spatial needs for the agency's efficient operation.

The result is an environment in which the simultaneous handling of seemingly antagonistic elements generates a spatial experience where the old and the new coexist harmoniously and with mutual respect. As the new elements form a sort of canvas or backdrop that allows the original architecture to express itself with greater force and clarity, smooth surfaces and clear colors were chosen to contrast markedly with the strong decorative statement made by the existing elements.

In terms of construction, all the home's original elements were painstakingly restored under the strict supervision of the National Institute of Fine Arts, while the new ones were meticulously incorporated and installed so as not to harm the existing structure. Thanks to this approach, the house can return to its original state without major complications or harmful consequences.

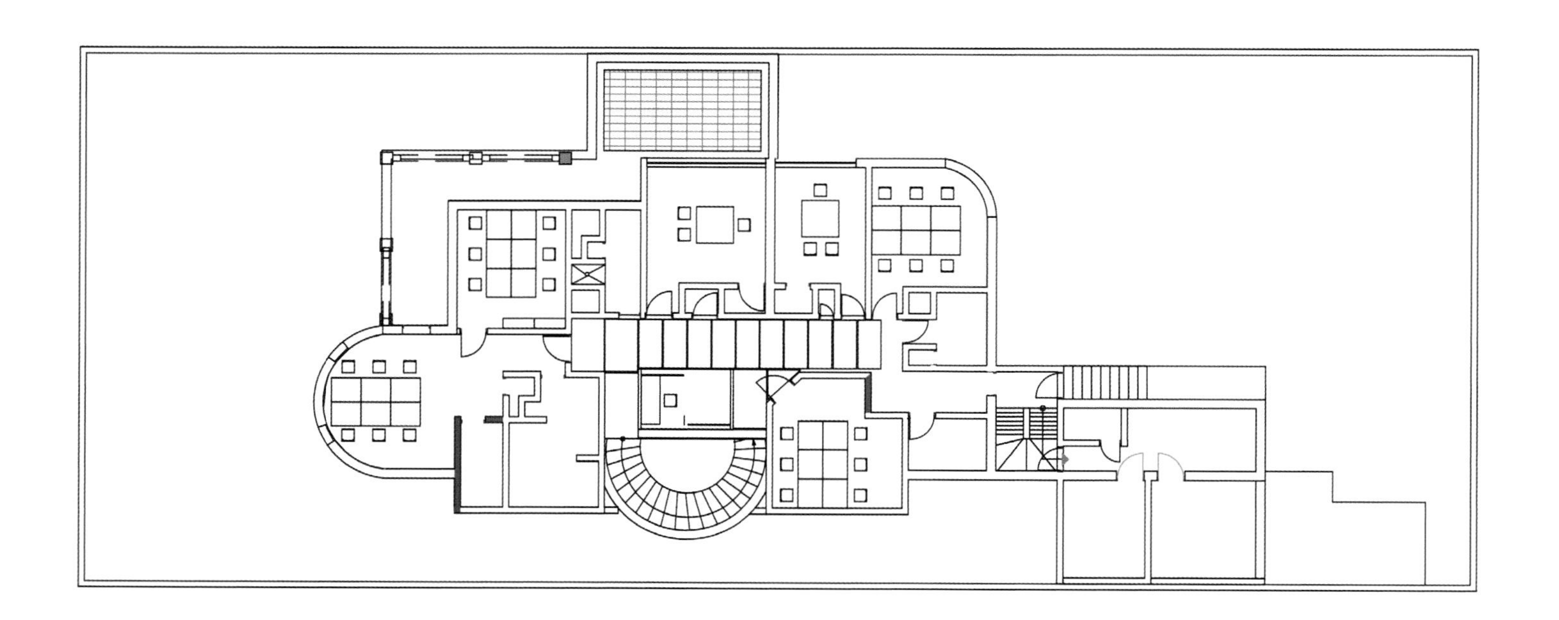

FIRST FLOOR PLAN

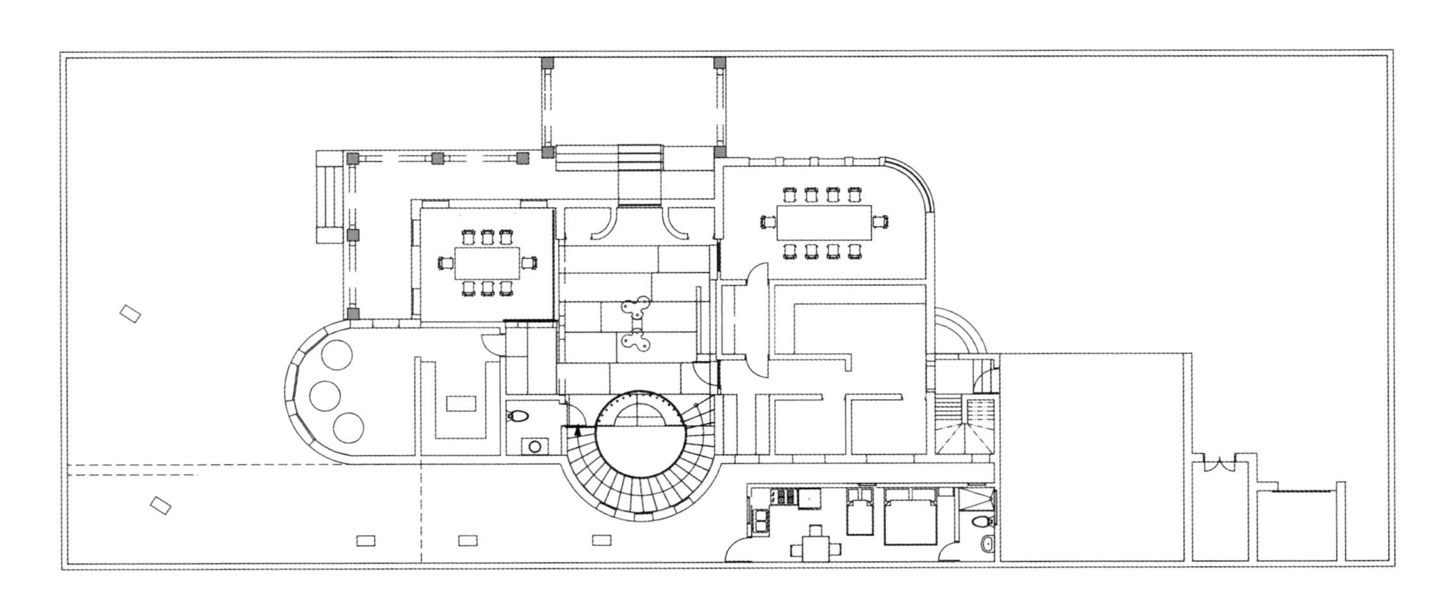

GROUND FLOOR PLAN

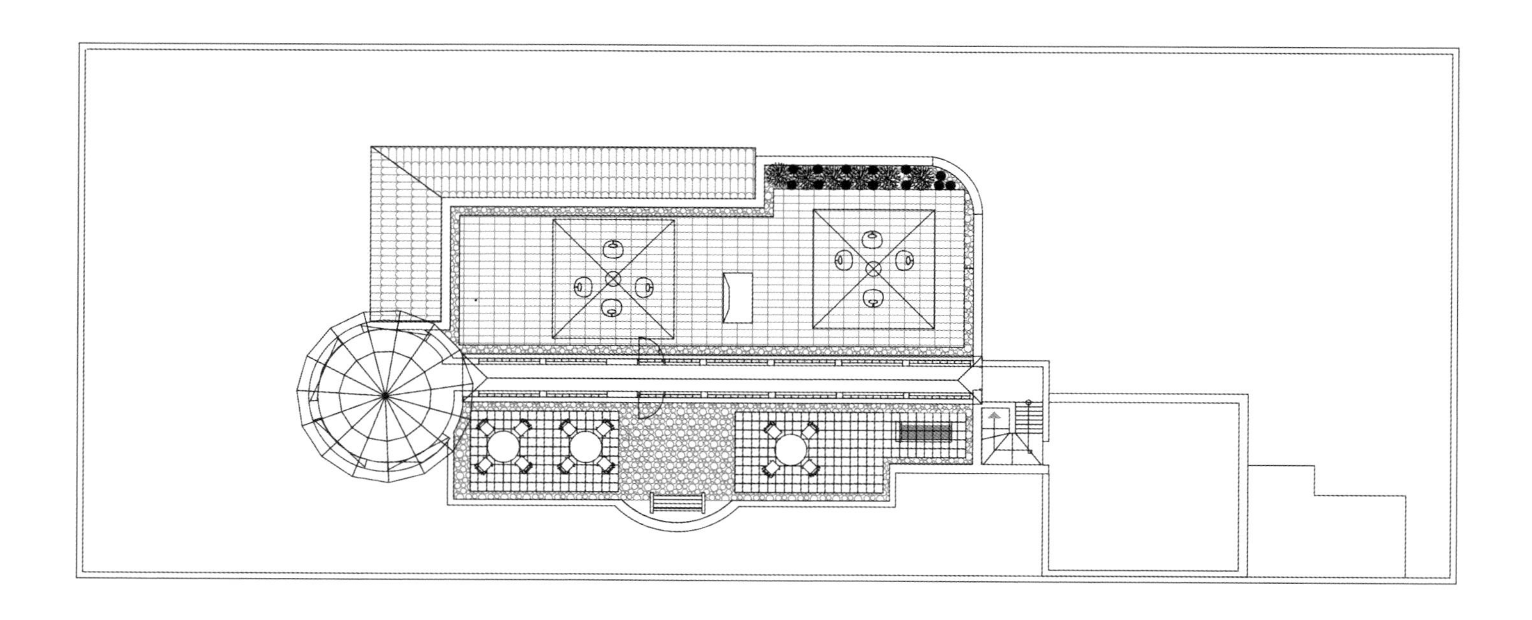

ROOF GARDEN

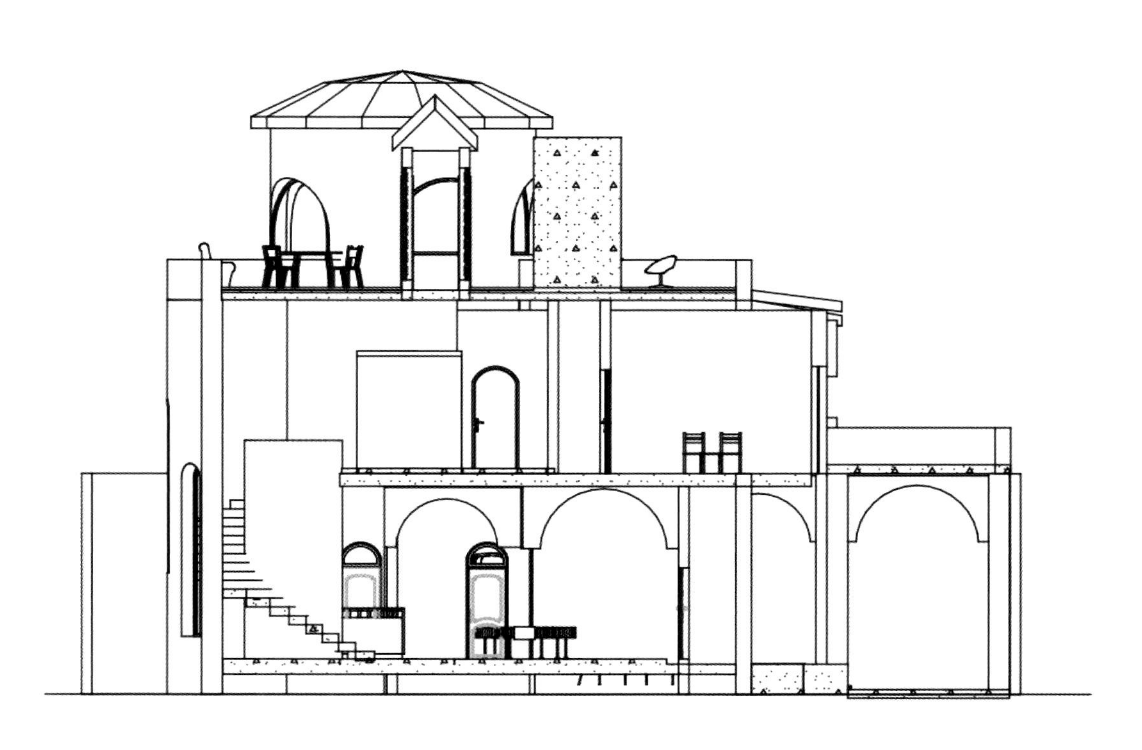

SECTION A-A

EMC Office

02

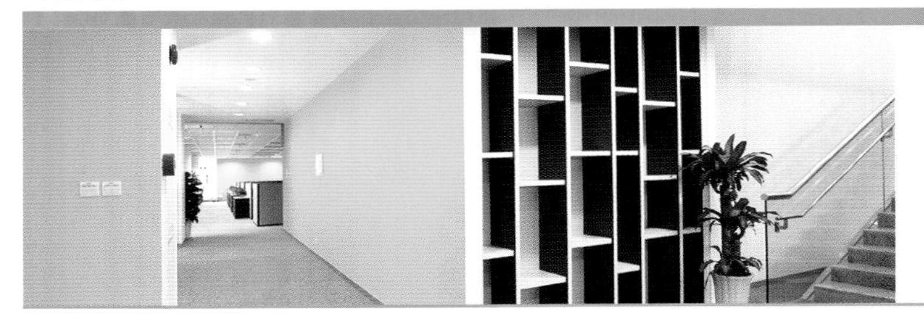

Design Company: DPWT Design Ltd

Project Location: Shanghai,China

Area: 8000 m^2

Photographer: Feng Zhiguang

The design of open office area mainly employs succinct tube light and lamp panel. The light reflecting on the bright-colored wall brings rich hierarchies for the office space and spices up the dull working environment. The tea room on the ground floor is in fact an activity room for the employees. The flash lamp and tube light on the false ceiling are hanging over the ping-pong table and painted glass walls, which has brought about enough light and stressed the bright colors. Going up the stairs and to the second floor, your eyes will be caught by the special false ceiling design: lamp standards with different heights shed irregular annular light which is reflected on the red leisure sofa, making the whole area a unique room for leisure. Many round droplights embellished over the bar are like stars in the night sky in the distance. At the end of the office area on the second floor, there is a gym specially built for the employees. Its droplight on the top is custom designed, which uses annular light trough with small tube light. Against the black painted ceiling, it adds a flavor of modernity and fashion to the gym.

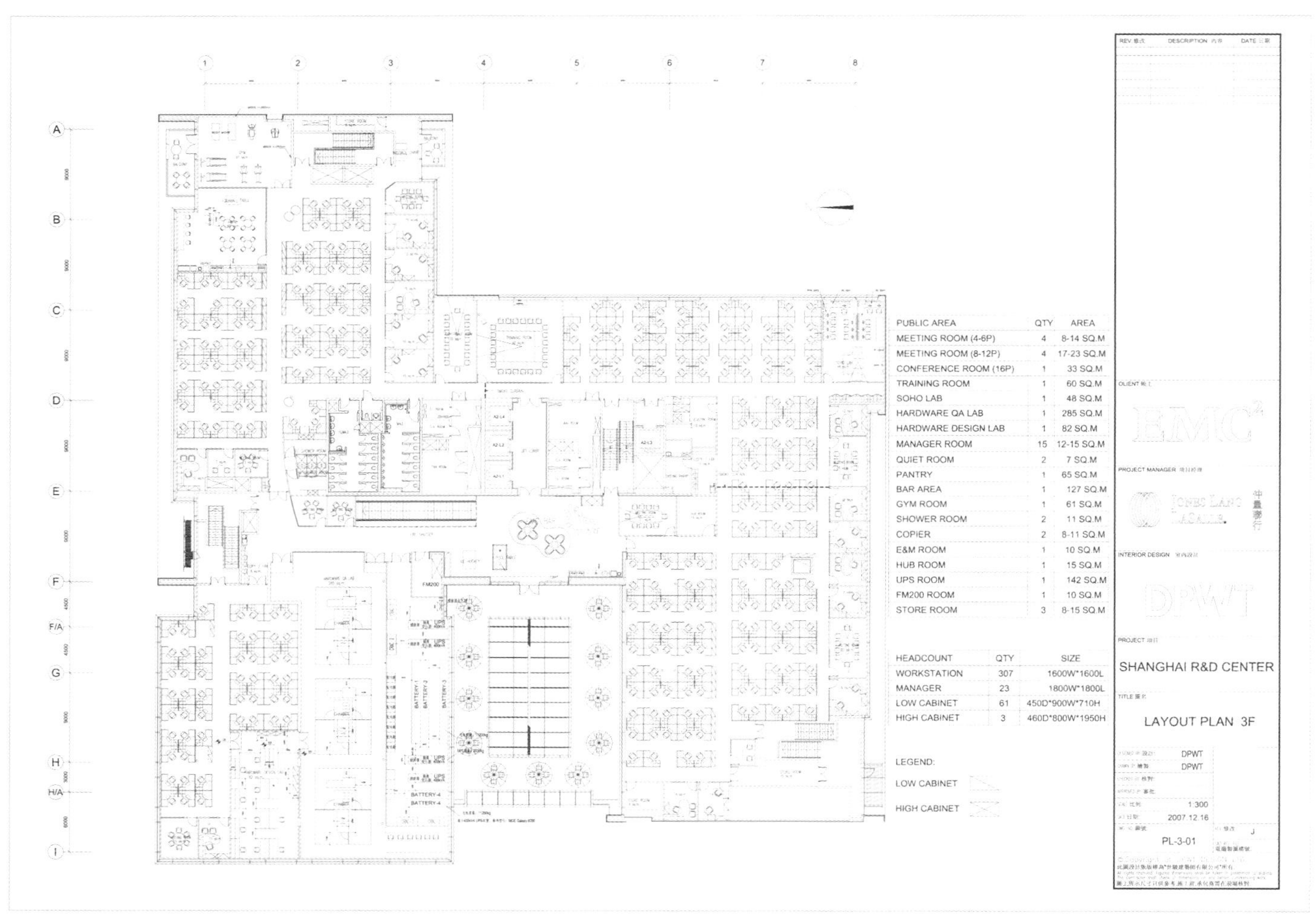
PUBLIC AREA | QTY | AREA
MEETING ROOM (4-6P) | 4 | 8-14 SQ.M
MEETING ROOM (8-12P) | 4 | 17-23 SQ.M
CONFERENCE ROOM (16P) | 1 | 33 SQ.M
TRAINING ROOM | 1 | 60 SQ.M
SOHO LAB | 1 | 48 SQ.M
HARDWARE QA LAB | 1 | 285 SQ.M
HARDWARE DESIGN LAB | 1 | 82 SQ.M
MANAGER ROOM | 15 | 12-15 SQ.M
QUIET ROOM | 2 | 7 SQ.M
PANTRY | 1 | 65 SQ.M
BAR AREA | 1 | 127 SQ.M
GYM ROOM | 1 | 61 SQ.M
SHOWER ROOM | 2 | 11 SQ.M
COPIER | 2 | 8-11 SQ.M
E&M ROOM | 1 | 10 SQ.M
HUB ROOM | 1 | 15 SQ.M
UPS ROOM | 1 | 142 SQ.M
FM200 ROOM | 1 | 10 SQ.M
STORE ROOM | 3 | 8-15 SQ.M
HEADCOUNT | QTY | SIZE
WORKSTATION | 307 | 1600W*1600L
MANAGER | 23 | 1800W*1800L
LOW CABINET | 61 | 450D*900W*710H
HIGH CABINET | 3 | 460D*800W*1950H
LEGEND:
LOW CABINET
HIGH CABINET
INTERIOR DESIGN
PROJECT
SHANGHAI R&D CENTER
TITLE
LAYOUT PLAN 3F
DPWT
DPWT
1:300
2007.12.16
PL-3-01
J

UL Guangzhou Office

03

Design Company: DPWT Design Ltd

Project Location:Guangzhou,China

Area:4000 m^2

Photographer: Chen Zhiwei

Walking into the reception area on the first floor, you are impressed with this fashionable and kempt workroom. The first thing that comes into your eyes is the simple and well-designed reception counter. It uses the common white color, but special materials of brushing lacquer and artificial stone, which presents a unique visual effect.

The glass and walls on both sides of the passageway to the experiment room are arc designed. Streaked window film fits with the wall design at the reception area.

On entering the second floor, we will be attracted by the training room at the entrance, which has employed a bold and innovative design instead of a traditional and serious one. The center part of the false ceiling is ordinary mineral wool board, which is rounded with plaster tablet for placing tube light. The height differentiation of two ceilings makes the hierarchies visually rich both in space and material. The design of tea cabinet wall stands out. The white tea cabinet is decorated with red and grey strips, and the wall is correspondingly clad in red and white painting glass, both of which are well designed. In addition, their collocation with dark-grey carpet and simple furniture makes the whole space lively and bright.

UL
Verification Services

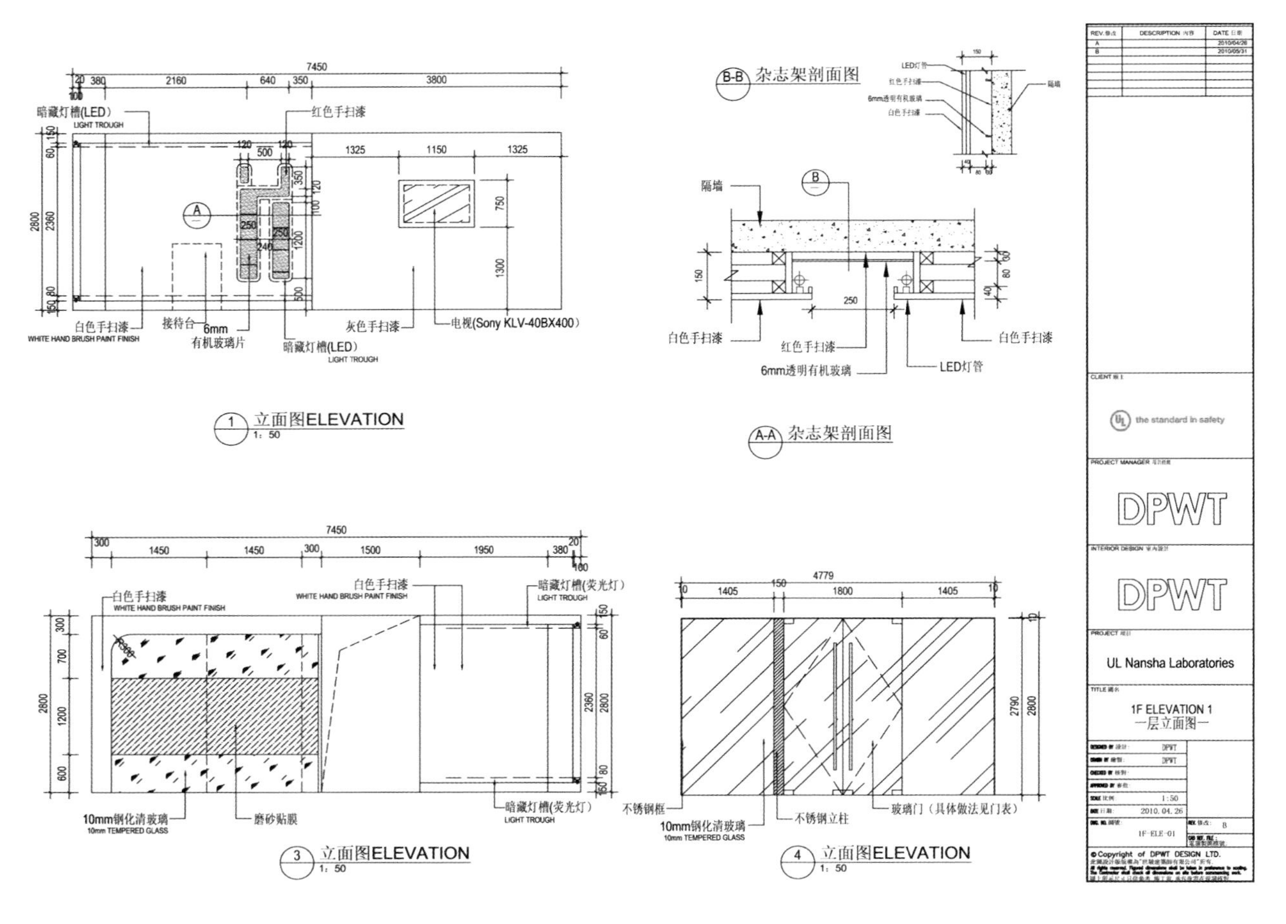

立面图ELEVATION
杂志架剖面图
UL Nansha Laboratories
1F ELEVATION 1
一层立面图
DPWT

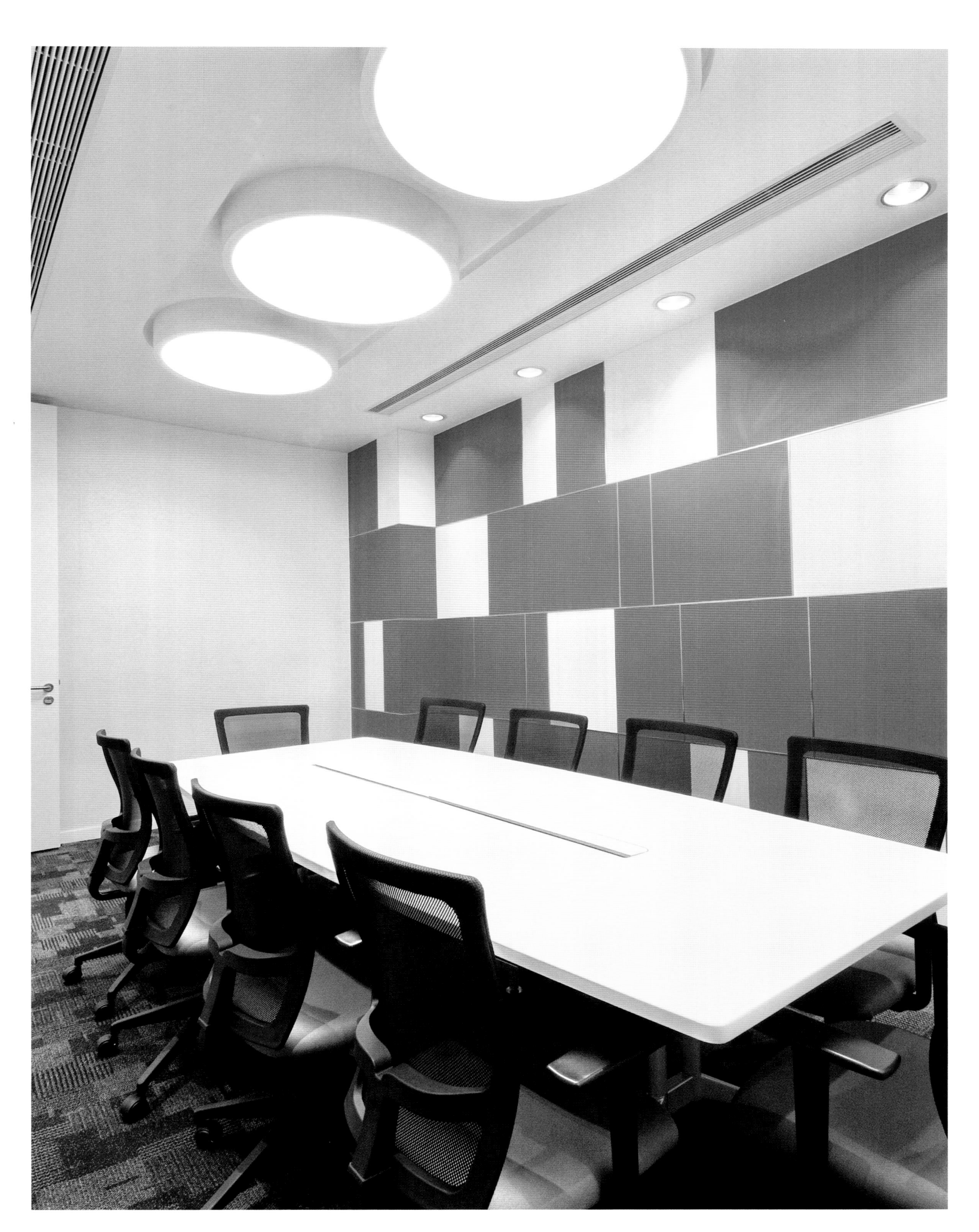

CONI ROVESCI

04

Design Company: Arch. DUILIO DAMILANO

Project Location: Cuneo

Area: 44,000 m^2

Photographer: Andrea Martiradonna

The offices you coin upside-down they constitute an appendix glass door that grafts him on the principal front of the shed of the new center of the DAMILANO GROUP s.r.l.
The volume breaks the rigid scheme of the industrial architecture projecting toward the outside the directional offices.
Idea is that to create, inside a grey and peripheral industrial context, an oasis in which to work in a pleasant and relaxed atmosphere.
For this motive the line drives some project you has been that to maintain a constant relationship with the outside and its disposition to vary of lights and shades and the idea to create one "operational greenhouse".
The offices are absorbed in a luxuriant vegetation, integral part of the project, that winds and he/she invites in the places of job.
The essences are cultivated with hydroponic techniques and they create upward climbing flows and descending downward. Horizontally the space is dominated instead by flows that they create sinuous volumes.
White ribbons run after him, they untie him and they fork creating salt reunions, platforms, lofts and offices. some rooms reunions are contained by upside-down cones, others from curtains, others still from green curtains.
From every angling of they open multiform foreshortenings, cuts vetrati and spatial interpenetration and views.
The sinuosity of the loft is taken back by the upside-down cones that, I suspended as stalactites, they channel downward the light.
From the office of the executive, two elevators, are possible to reach a further oasis drawn on the roof of the offices as an extreme shelter through.
The white flow is interrupted by the cone anthracite that concludes the whirling of the forms and it leans out on the area meeting. This last is absorbed in the green and constituted by a platform in wood inside which him trove the center of the project: the man and the comfort, the Ing and Yang.
The precariousness of this state of equilibrium is underlined by the lightness and instability of a floor of leaves...

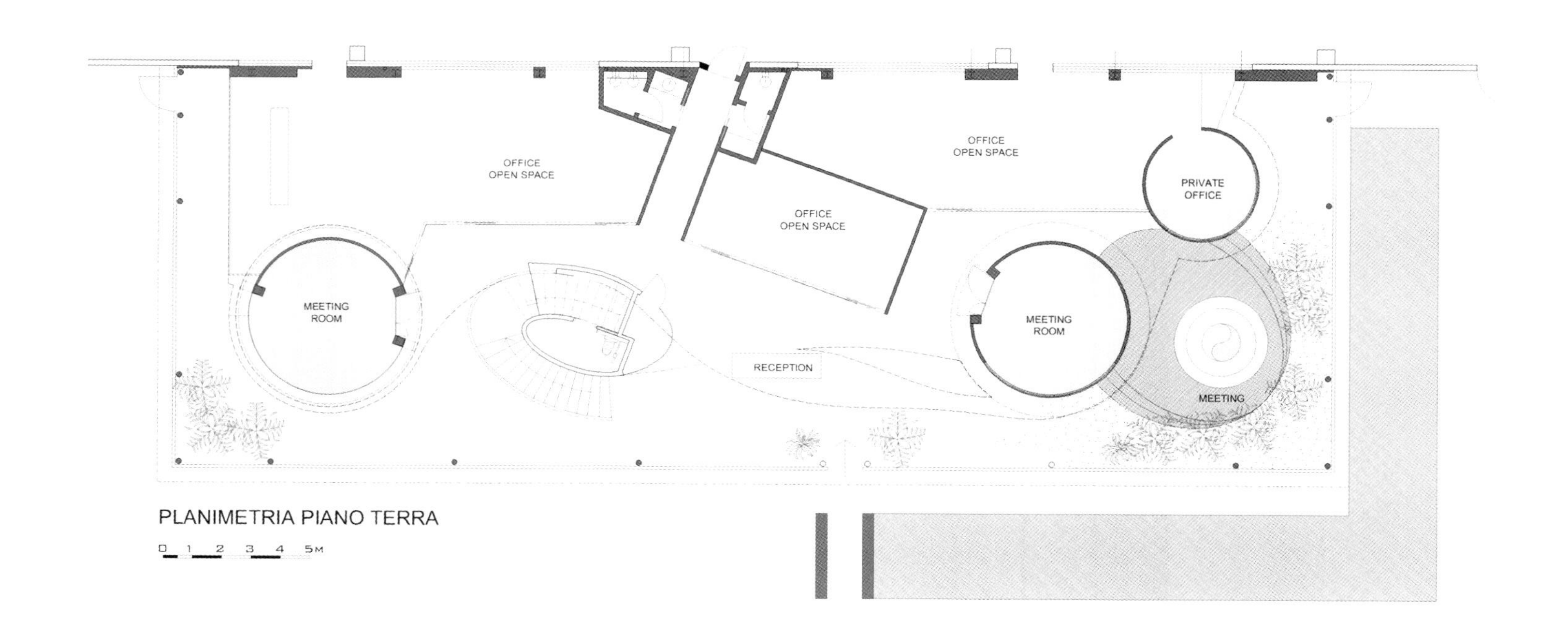

PLANIMETRIA PIANO TERRA

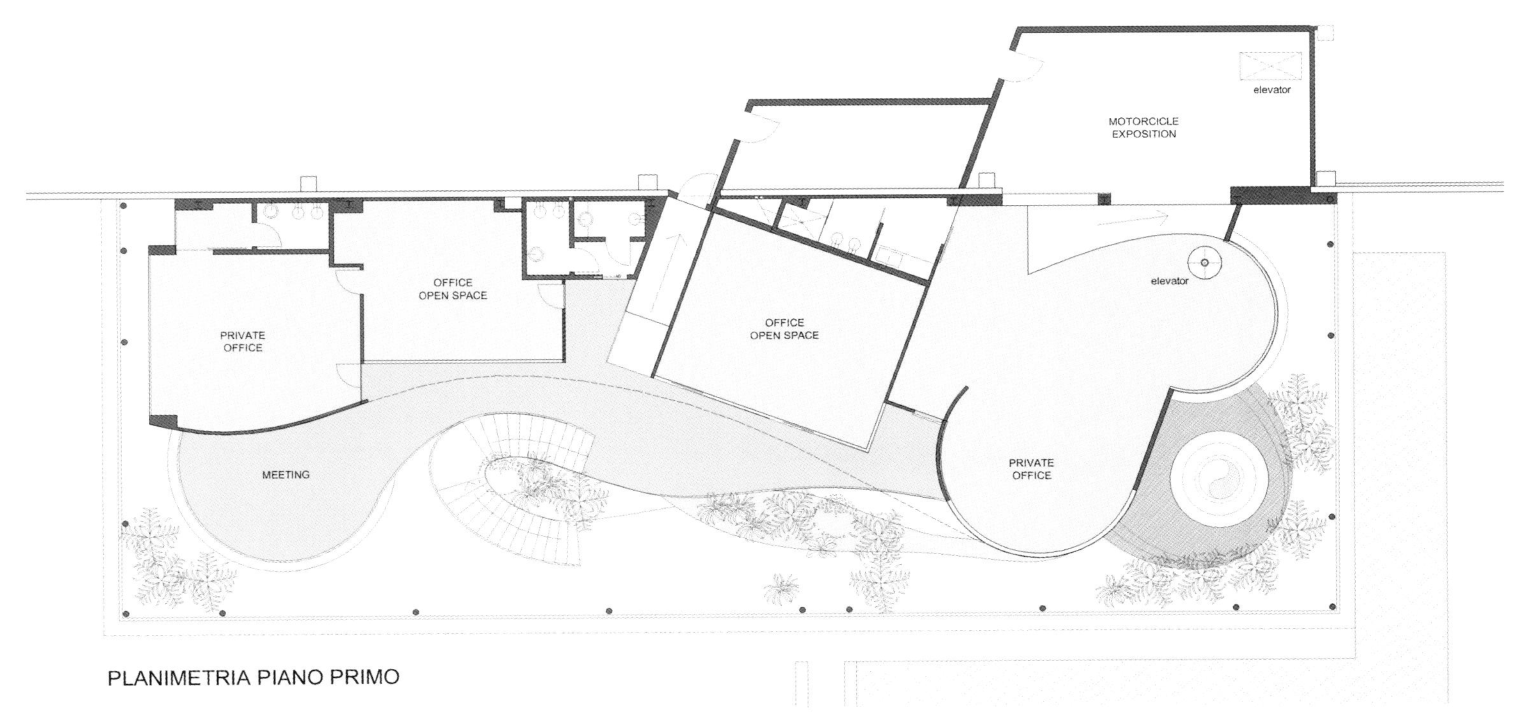

PLANIMETRIA PIANO PRIMO

Industrial Association

05

Design Company: ISA STEIN Studio
Project Location: Linz, Austria
Area: 550 m²
Photographer: ISA STEIN

The "Industrial Association" did want to have a more showing appearance. To point out the industrial theme we took metal as a basis material in most areas. You can find it in the furniture, doors, staircase and cladding. In addition we worked with quite natural colors like white and cream and the material glass.
As it is a renovation of a 20 year old building, we broke down the windows, in order to create more transparency and space. We opened the cladding and added the cube to achieve full transparency and strengthen the outside appearance. We also designed a balcony as an extension of the inner space.
For the recognition value the "IV"-Logo is written on the cladding. On the ground floor little metal "IV"-Logos are hanging from the ceiling for the pedestrians to see.
As the three floors of the Industrial Association should be a connected space, the interior staircase heads one up to the big conference room. There we again worked with a lot of transparency, so that we get the light shinning from east to west. We also put an accent to the great view onto the Pöstlingberg, an emblem of Linz.
The steering committee room is next to the "come together" area on the ground floor and separated by a partition wall. In case of a big event it is possible to interconnect the two areas and transfer it into one big foyer. Behind the parked partition wall is enough room to store the entire interior from the steering committee room.
In the pre zone of the first floor, there is a lounge area with Moroso-seating on a cow coat, as well as felt lampshades.
The internal offices are located in the second floor and are also designed as an open floor plan. Only furniture is dividing the space. For acoustic reasons we added area covered curtains.
At the entrance zone of the office space on the third floor an image and a meeting point with an integrated bench is welcoming you.

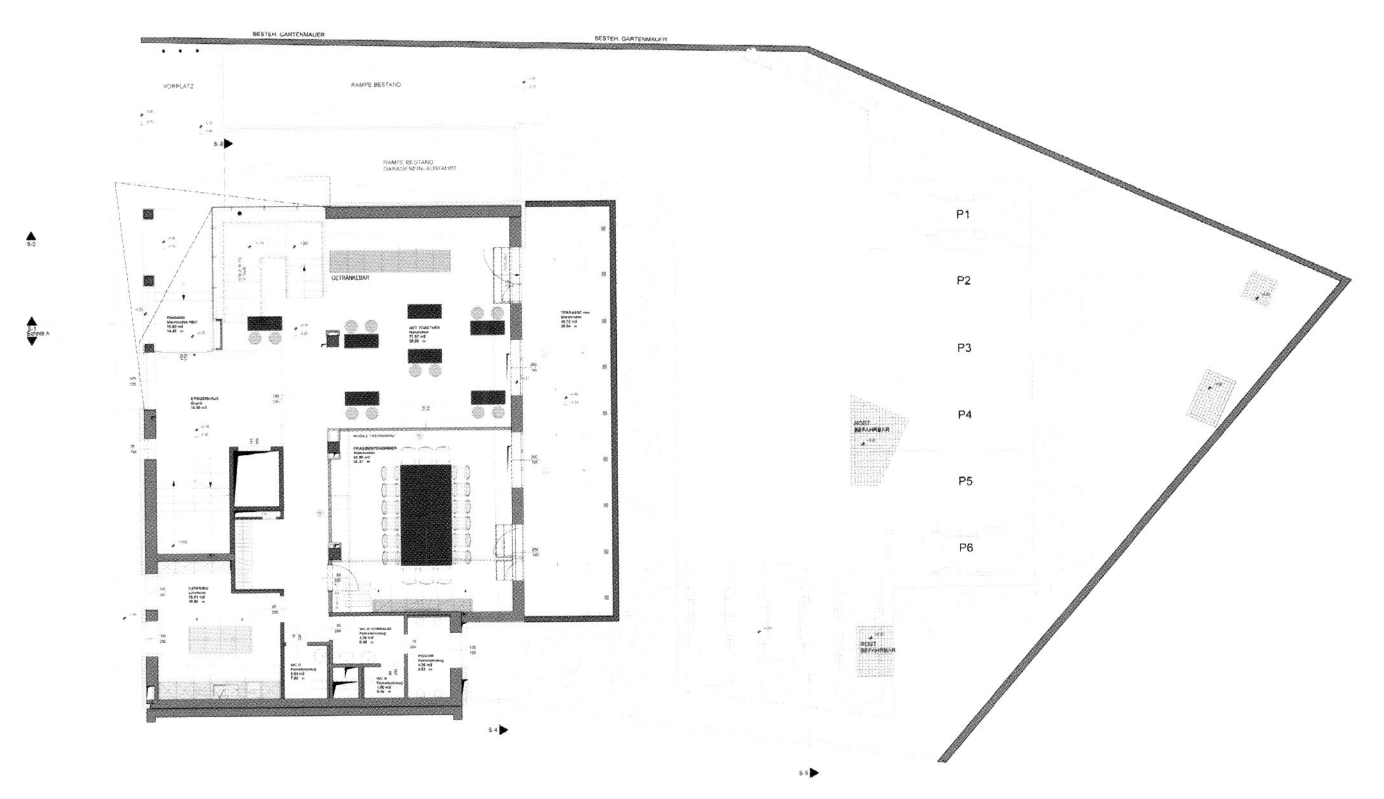
BESTEH. GARTENMAUER
BESTEH. GARTENMAUER
VORPLATZ
RAMPE BESTAND
GETRÄNKEBAR
P1
P2
P3
P4
P5
P6
ROST BEFAHRBAR
ROST BEFAHRBAR

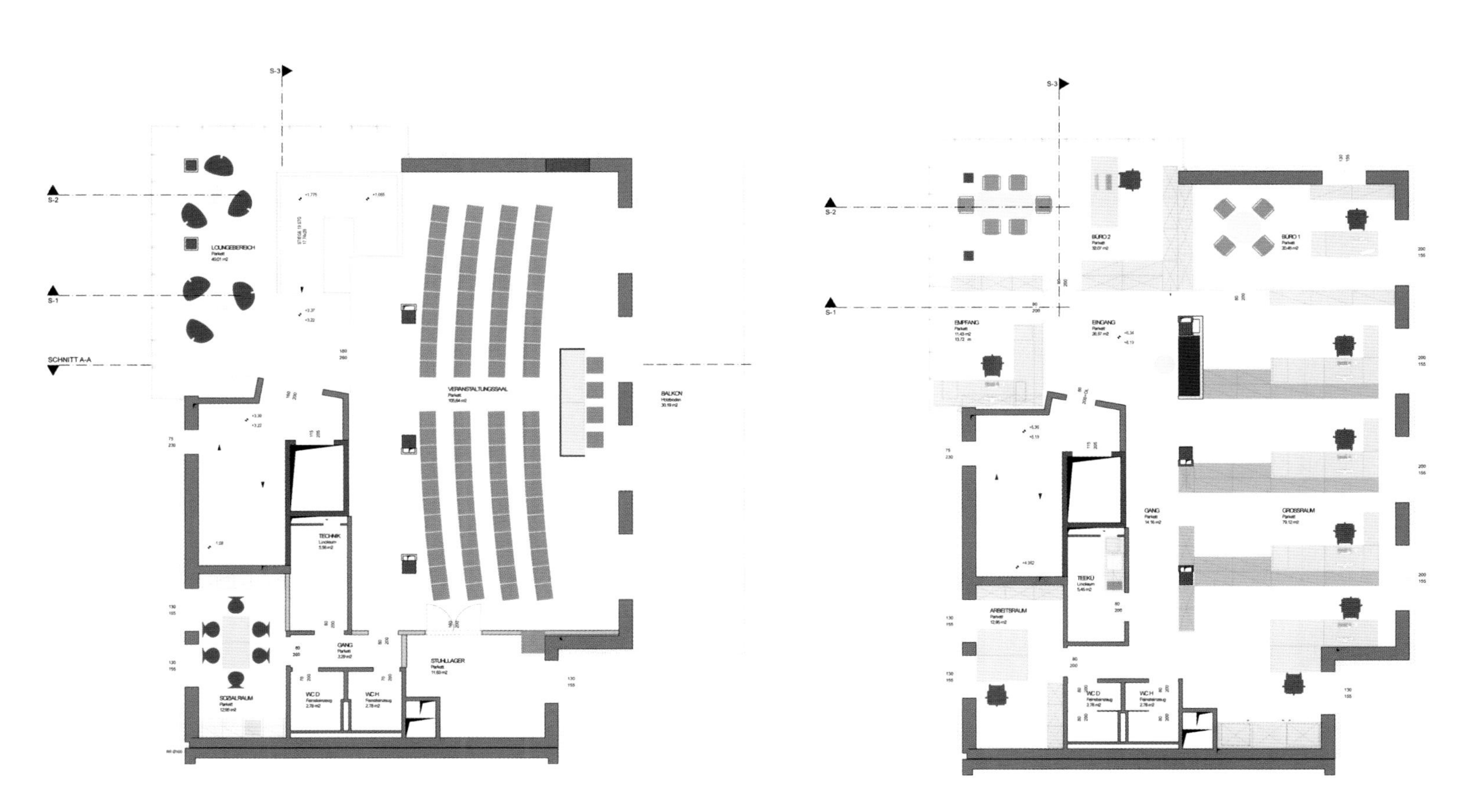
S-3
S-2
S-1
SCHNITT A-A
VERANSTALTUNGSSAAL
BALKON
STUHLLAGER
TECHNIK
GANG
S-3
S-2
S-1
EMPFANG
EINGANG
GANG
GROSSRAUM
ARBEITSRAUM

iv
Industrieland
Oberösterreich
iv

Offen für Menschen
und Neues.
Wir erzeugen Zukunft
Die OÖ. Industrie

The Long Barn Studio

06

Design Company: Nicolas Tye Architects
Project Location: Maulden, Bedfordshire, MK45 2EA, UK
Area: 220 m²
Photographer: Nerida Howard

The concept of the design is based around an elegant glass, rectilinear box which is enclosed at both ends with larch clad 'book-ends'. The larch is treated so weathering appears uniformed both externally and internally creating seamless lines. The 'frame less' glass panels allow high levels of natural daylight into the studio along the northern side and in addition allows wide views of the surrounding landscape to become integrated into the studio environment. Along the southern elevation the larch clad timber pods also punctuate the glass facades, which remove issues with overheating. Each pod has a dedicated use including an Architectural Library, reprographics area, toilets and meeting room. Cor-ten steel detailing is also used throughout the studio to reflect the original agricultural nature of the site and accentuates the smaller windows and linear roofline. Within the studio's interior a continuous limestone floor enhances the main axis, and the space is broken up through a series of wenge 'pods' which contain ancillary elements including the kitchen, storage containing a materials and document library, and individual staff storage.

north elevation (1:100 scale when printed at A3)

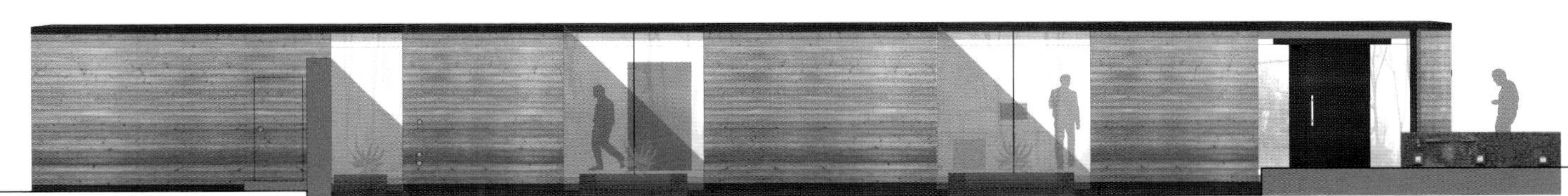

south elevation (1:100 scale when printed at A3)

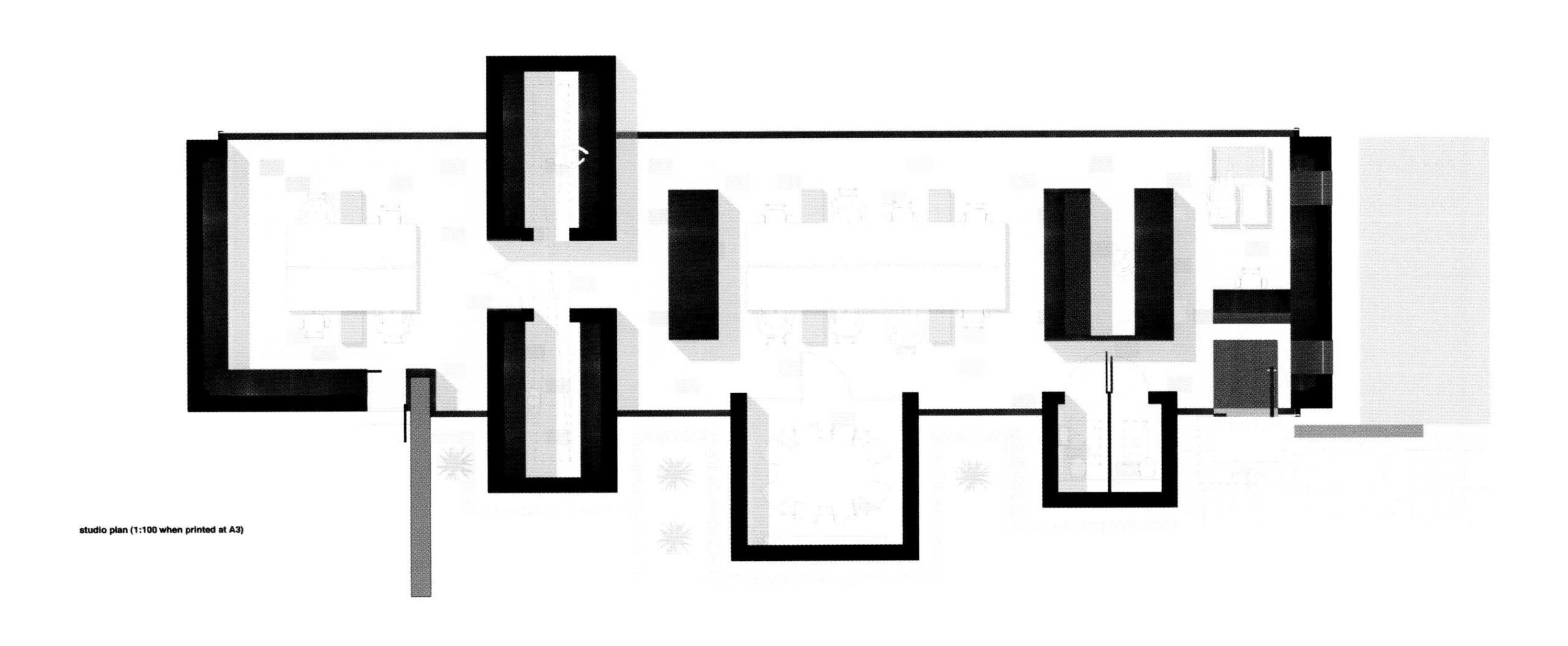

studio plan (1:100 when printed at A3)

BCP Offices

07

Design Company: Arch. José Orrego

Project Location: CRONOS Building, Lima, Perú

photographer: Juan Solano

A brand new concept was developed for one of the most important Banks in Perú.
Contemporary work requires new adjustable spaces, which were the center of the research made by our design team.
Based on that, the project tries to optimize the system, interaction and work team: every working spot was integrated to
each other so that a central multi-proposal space would appear.
The intention at BCP was to reduce the ocupation ratio without neglecting the quiality of the office environment.
The concept of lighting includes management systems that sinchronize artificial and natural light at the same time to create a comfortable atmosphere.
The design bet on inner perspectives to gain visual range and amplitude.
The project is supposed to recreate the sensation of a Urban Café in an office context.

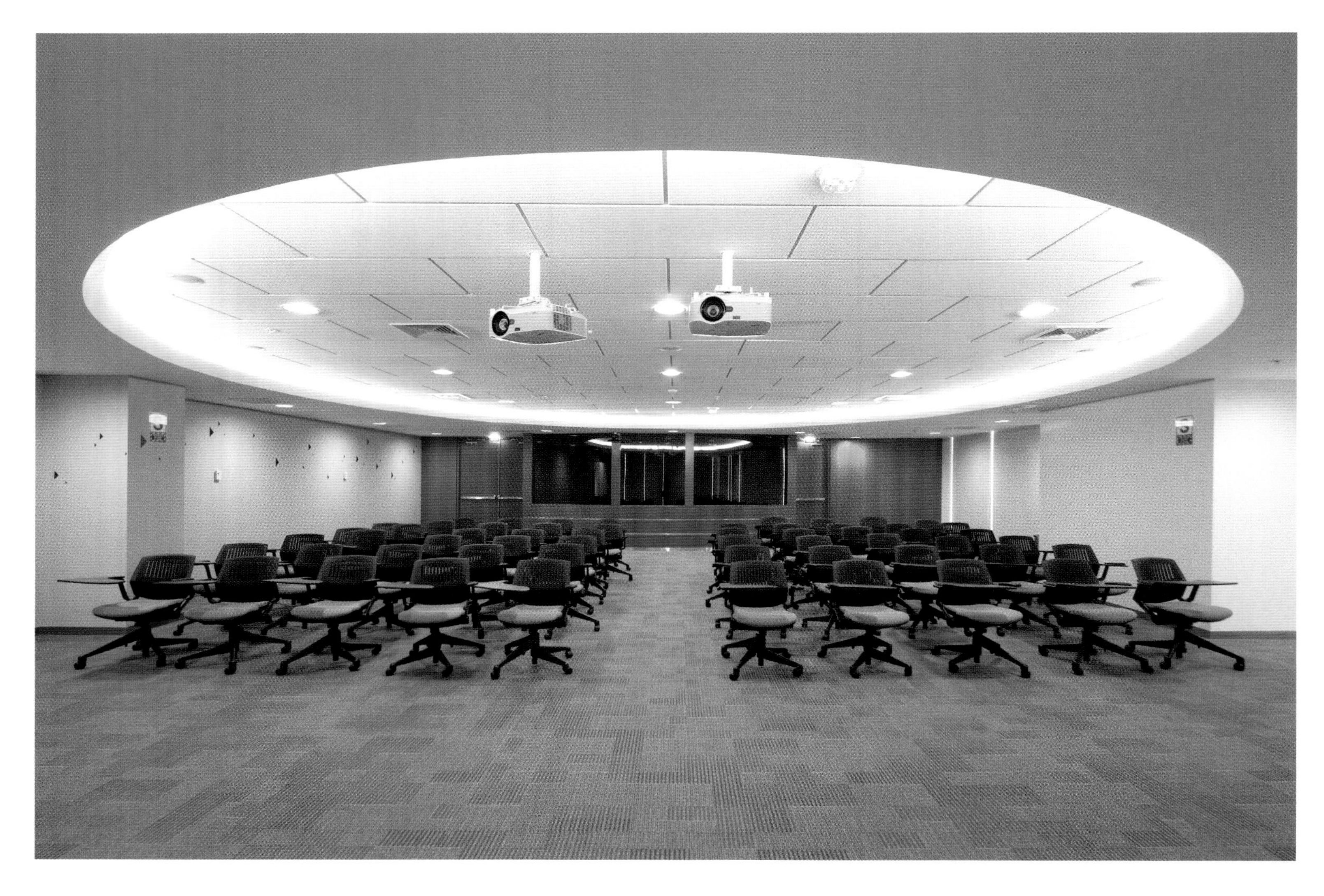

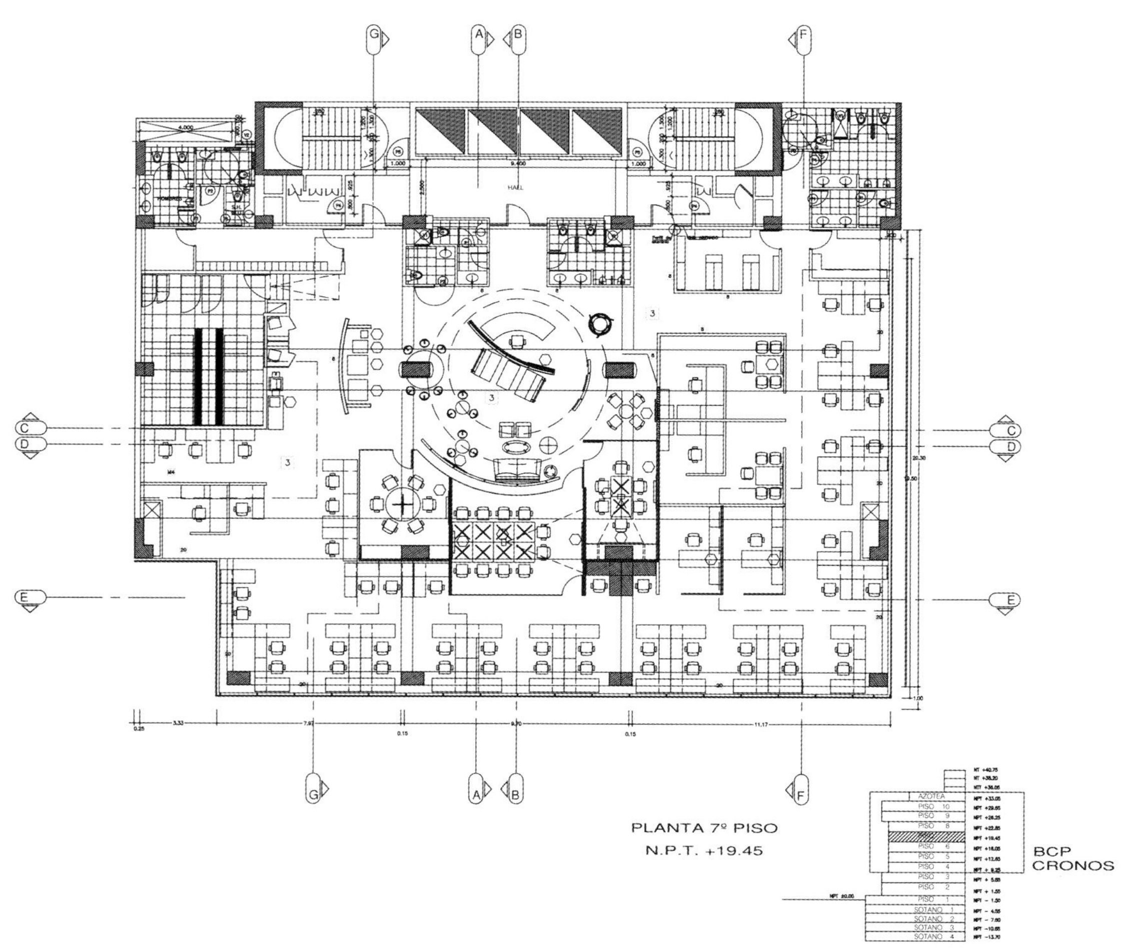
PLANTA 7º PISO
N.P.T. +19.45
BCP
CRONOS

Invescon Offices

08

Design Company: Agustí Costa Design Studio

Project Location: Solsona - Spain

Area : 105 m²

Photographer: David Cardelús

Estate agent's in Solsona, located on the premises of a new building. Solsona is a small town with a huge architectural heritage, but with an essence of country, forest and rural spirit, all mixed with an enterprising way of doing things. There is a school of Agricultural Education and it is the headquarters of the Technological and Forest Centre in Catalonia with 150 investigators. There is also mechanical industry, construction industry and manufacturing of country products.

The design project wants to reflect all this as well as the estate agent's activity of the company, opposing to a totally contemporary and neutral atmosphere, an untreated pine wood cover, a ceramic block face made of thermoclay and a second one made of coconut carpet, everything with the same aim of evoking the forest, the country and the housing construction and of giving the necessary warmth to the employees in an atmosphere which is close to the imaginary group of people in Solsona.

The main element is the false ceiling of wooden strips 4.5 cm wide and 6 of high, separated from each also 4.5 cm. Transversely, the office is modulated depending of these 4.5 cm. Inside this open ceiling, are accommodated the guides of sliding glass doors, air-conditioning tubes that distribute air through the ceiling grid, and thus, avoid the direct impact, the strips of fluorescent T-5, etc. The staircase is a volume permeable, located in the area of double height formed with the same wooden slats.

The furniture, designed entirely for the occasion, consists of tables, base cabinets, countertops and the large L-shaped cabinet it is laminate of grey colour, which has a slight silver hue. The same material is also used to build the screens on both floors.

The showcase is a light volume, a large glass lamp resting on the ground and that performs three functions. On the inside have open compartments to hold samples of finished products, also makes seat for three people, and in the outside, their inclined planes allow to place ads propaganda

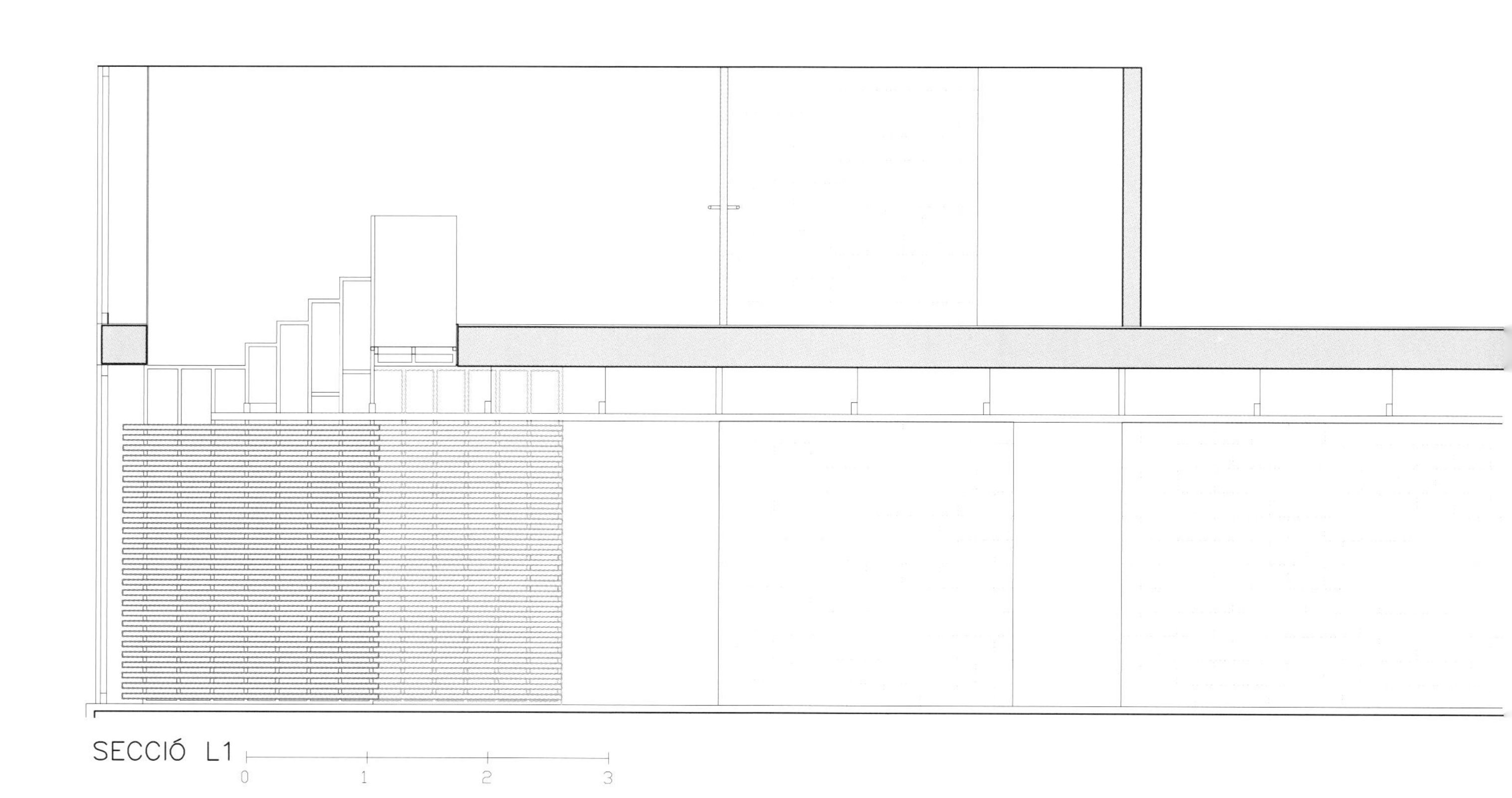
SECCIÓ L1
0
1
2
3

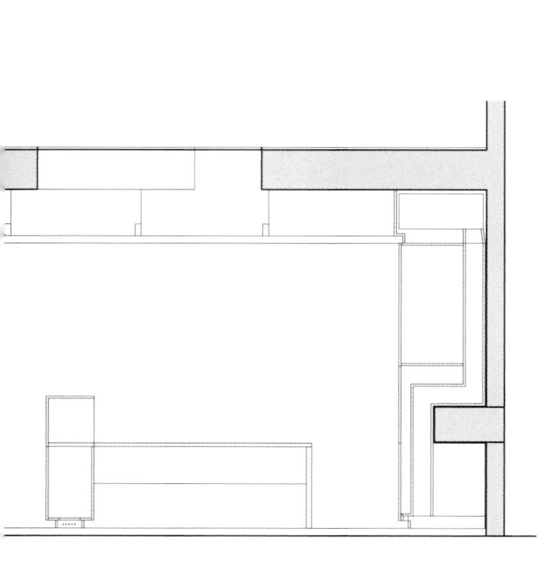

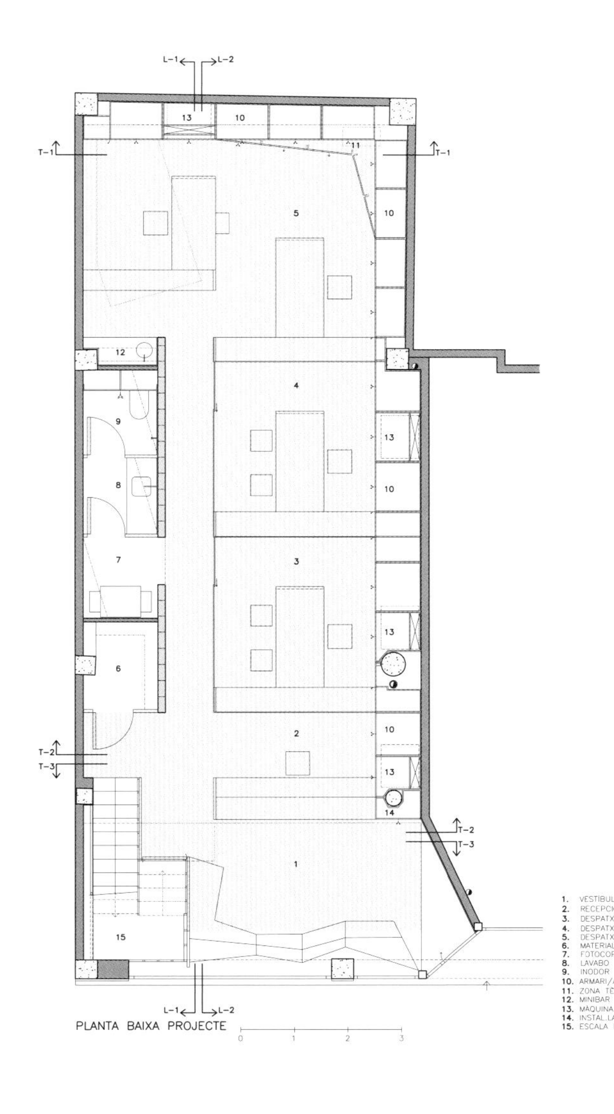

1. VESTÍBUL/APARADOR
2. RECEPCIÓ
3. DESPATX ATENCIÓ AL PÚBLIC-1
4. DESPATX ATENCIÓ AL PÚBLIC-2
5. DESPATX-3
6. MATERIAL
7. FOTOCOPIADORA
8. LAVABO
9. INODOR
10. ARMARI/ARXIU
11. ZONA TÈCNICA
12. MINIBAR
13. MÀQUINA CLIMATITZACIÓ
14. INSTAL.LACIONS
15. ESCALA ENTRESOLAT

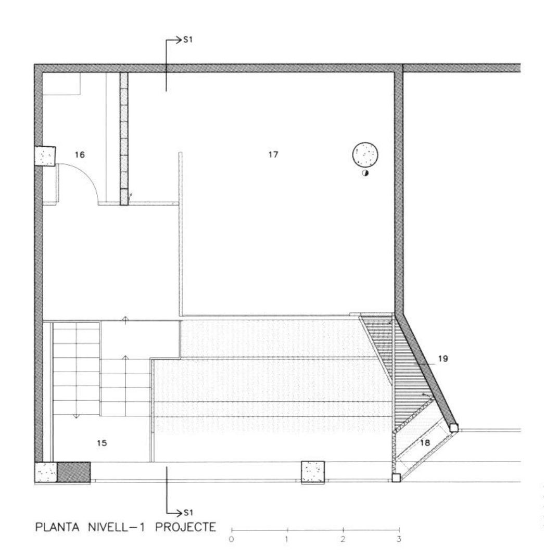

15. ESCALA ENTRESOLAT
16. MATERIAL INFORMÀTICA
17. ESPAI VACANT (FUTUR DESPATX DIRECCIÓ)
18. MÀQUINA EXTERIOR CLIMATITZACIÓ
19. PASSERA DE SERVEI

Opman Offices

09

Design Company: Agustí Costa Design Studio

Project Location: Manresa (Barcelona) - Spain

Area : 123 m²

Photographer: David Cardelús

It is an offices designed for public works engineering, located on the premises of a building with a classical air in its frontage, which was built in the early forties. The design project consists in totally remodelling the current office, suggesting open spaces, transformable by means of divisions which are electronically activated and mobile pieces of furniture, making it easier to create the necessary kind of space for each moment, extending or reducing the surface of the work space at will. The entrance and the meeting room can also have different degrees of visual communication with the rest of the place by means of some sliding glass sheets and mobile folding screens. Some antique walls are covered with glass and there are some glass walls which increase the feeling of spaciousness and interact with the pre-existence, such as the communal skylight of the building.

The main elements of the work room are a fixed table of 13.20m in length and a double-sided metal shelving, also fixed and 6.30m long. The table, supported by 4 drawers, the network installation has been hidden and the workplaces are illuminated with natural light from the street, and in the evening, with the cascade of fluorescent continuous light that hides in the hole perimeter of the ceiling, which also have the emergency, music and other hidden utilities. The metal shelving incorporates the broadcast air-conditioned, the shelf is double-sided and in the background, between the two shelves, there is a translucent glass separator, in order to provide the diffusion of natural light from the street and the communal skylight. The remaining furniture, consists of furniture drawers and small table, with wheels, deals the middle of the room and can take various combinations, with the help of two electric white screens, with the roll hidden in the ceiling, allows have the space required for each occasion.

A final decision has been use the natural light in order to get one of the main objectives set at the beginning: fully exploit their potential to transmit the light in all directions. Therefore, it has built in glass all that was susceptible to make with glass: the separation of big room with the meeting room, the lining the back wall of the corridor, the separation between the two shelves, the volume that makes up the room waiting, all sliding doors and finally the lining of the walls of the communal skylight, the waiting room and the access volume to the communal skylight. These latter, with optical glass insulation, translucent and without profiles, equivalent to a kind of upholstered of glass placed on the empty and the full of old wall, and let filter in the light of the communal skylight in more or less intensity depending on the situation of old opening, the time of day or weather.

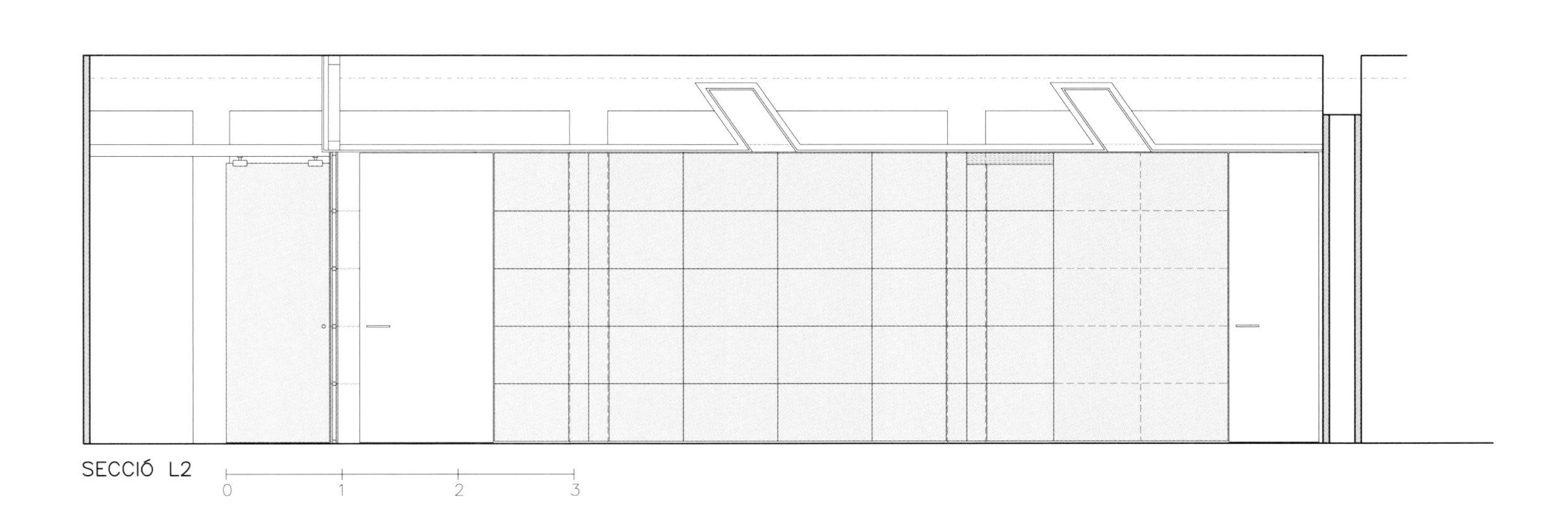
SECCIÓ L2
0
1
2
3

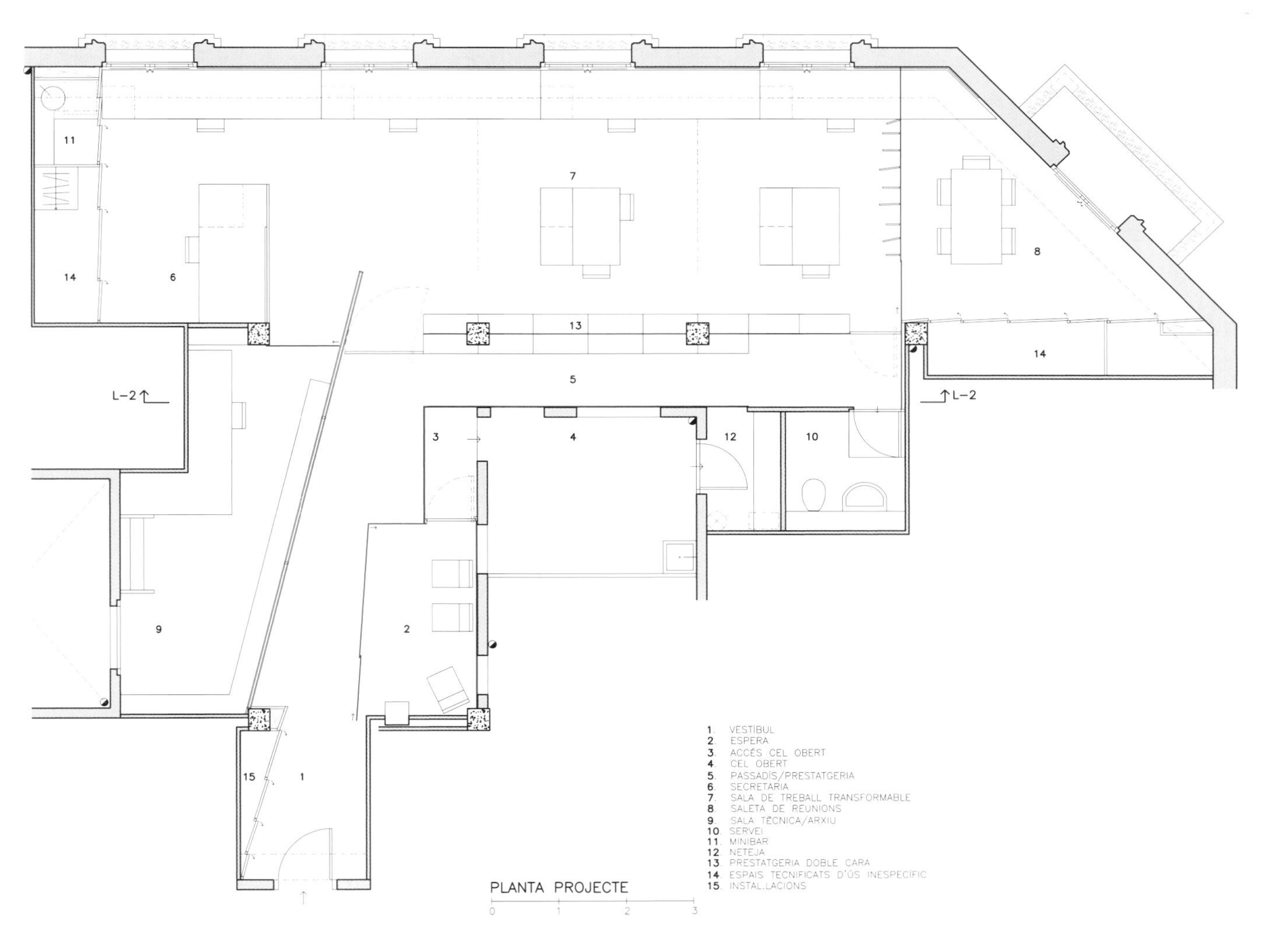
L-2
L-2
1. VESTÍBUL
2. ESPERA
3. ACCÉS CEL OBERT
4. CEL OBERT
5. PASSADÍS/PRESTATGERIA
6. SECRETARIA
7. SALA DE TREBALL TRANSFORMABLE
8. SALETA DE REUNIONS
9. SALA TÈCNICA/ARXIU
10. SERVEI
11. MINIBAR
12. NETEJA
13. PRESTATGERIA DOBLE CARA
14. ESPAIS TECNIFICATS D'ÚS INESPECÍFIC
15. INSTAL.LACIONS
PLANTA PROJECTE
0 1 2 3

Gonzalo Mardones Viviani Arquitectos Studio

10

Design Company: Gonzalo Mardones Viviani

Project Location: Av. del Valle 869, of 01, Huechuraba, Santiago de Chile.

Area : 500 m^2

Photographer: Guy Wenborne

As architecture studio we were attracted with the idea to work in the Vanguard Building, our last Offices project. This building, as all our works was created with only one material, in this opportunity with white steal. This material was used in all exteriors and interiors of the building, also in the six storey hall: the soul of the building where the white steal was neutral to permit the light and the total space were protagonist.
As the same way our studio was entirely white. Place in the first sub floor was an L shaped turn a patio of 11 x 11 mts that is the extension of our atelier and offices. Located in the first sub floor (in a 500 mts place) permit us to project our studio in a space with an interior height of 330 cms with very success natural light. The plan was not orthogonal, that permit as to tense the circulations and the space.
The main idea of our studio project was work in the better spatial and natural light conditions.
At the same way that all our projects and the same way of the Vanguard Offices Building too we work with one material and one colour, we opted for the white: all the walls and roofs where white. The floor was a light grey ceramic that collaborate with the white sensation (the same floor in the entire studio) and we separate the spaces with transparent crystals that collaborate too. The white permit light and space is the protagonist. The furniture was white too and also the chairs were white, there are no elements that compete with the white, finally with the white we permit the light was the most important material.

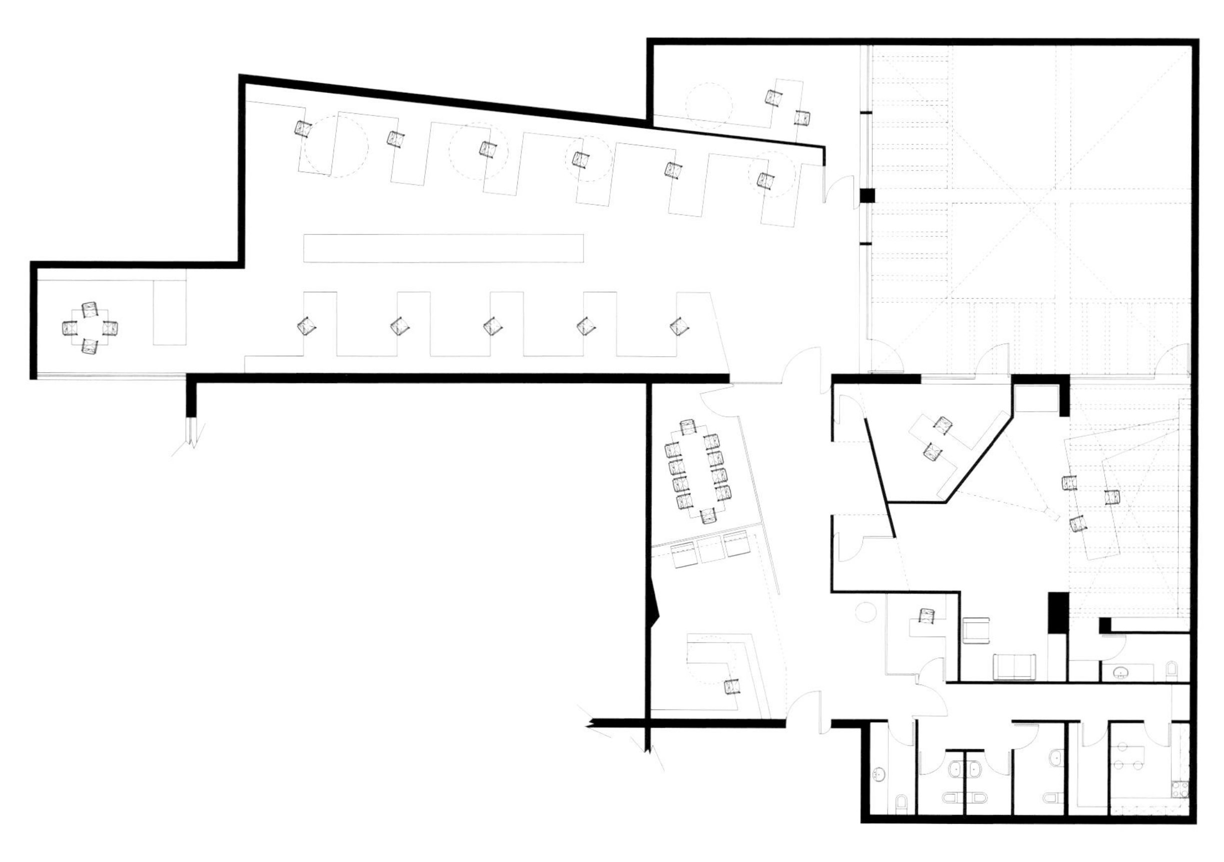

Office Herengracht, Amsterdam, the Netherlands

11

Design Company: i29 I interior architects / eckhardt&leeuwenstein architects
Project Location: Herengracht, Amsterdam The Netherlands
Area: 240 m^2
Photographer: i29 I interior architects

The board of an investment group in capital stock, wanted to have a self-called 'power office'. i29 I interior architects and Eckhardt&Leeuwenstein, two offices which collaborated during this project, created this by placing every board member in the spotlight on a playful way.

All three boardrooms and a lounge are executed in an overall design concept. Large round lampshades, spray painted gold on the inside, seem to cast light and shadow oval marks throughout the whole space. By this, a playful pattern of golden ovals contrasts with the angular cabinets and desks, which are executed in black stained ash wood. In the flooring the oval shaped forms continue by using light and dark grey carpet. Also, these ovals define the separate working areas.

The lounge area has, in combination with the white marble flooring these same light/shadow patterns that cover the bar and benches in silver fabrics. This area can be used for presentations or social working, with an integrated flat screen in the bar and data connections in all pieces of furniture. The existing space is set in a 17th century historic building, at one of the most famous canals of Amsterdam called 'de gouden bocht'. All existing ornaments and details are painted white.

The keynotes for this company are money and power. The design concept expresses this by setting all members of the board literally in the spotlights. The golden and silver ovals shatter through the spaces like golden coins. Where it is all about.., in investment and stock trading.

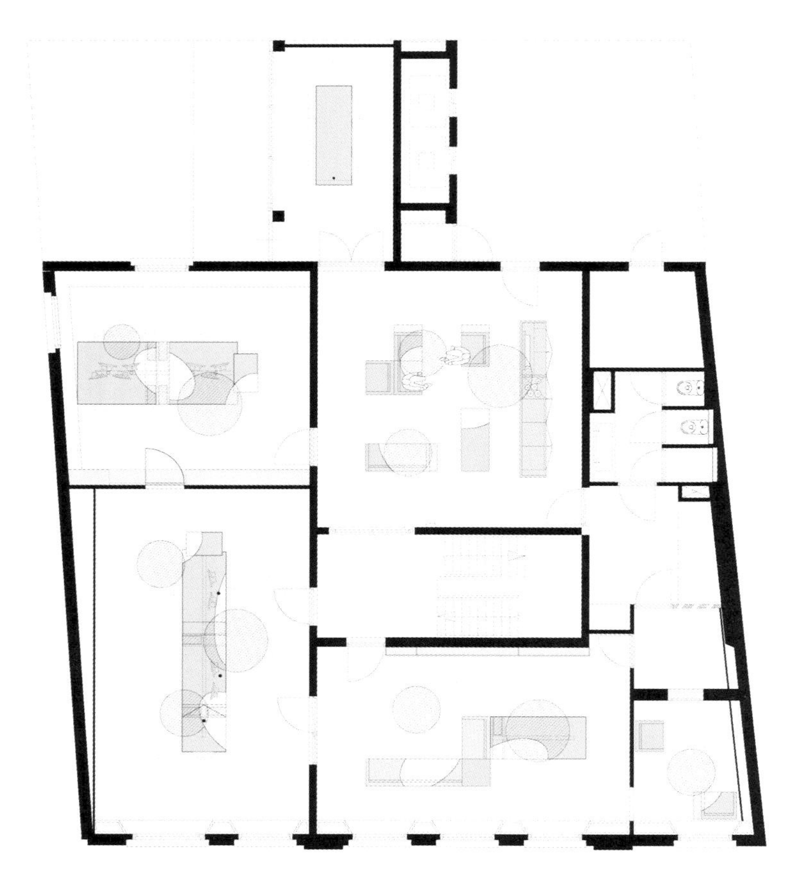

Luminare

12

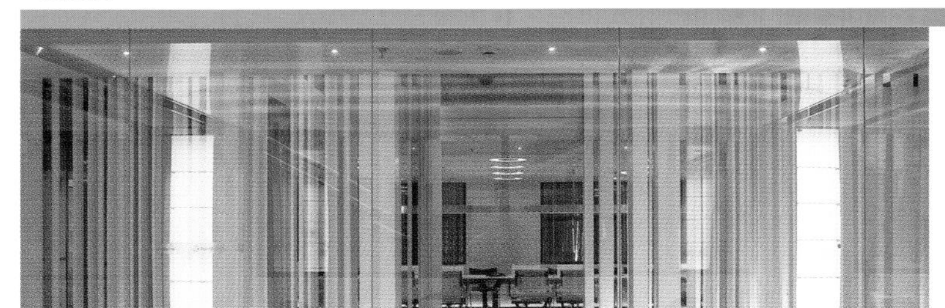

Design Company: Hank M. Chao / MoHen Design International

Area: 1200 m^2

Photographer: MoHen Design International/Maoder Chou

This office located at Shanghai city center Jing An Temple area, it belongs to those old factory renovated into new creative industry park office. Client professions in LED marketing, this part baffle me. After all, I am lacking the experience of using LED lights to accomplish the whole illuminate plan, do not have any concept of using how many numbers of light. Therefore, the rising enthusiasm has been reduced into half. If we look at the architecture pattern with plane surface and section, one roughly six meter rectangular lifted high division, need to have some creative and interesting method to pitching in to leave no regret. After think it in many ways, finally decide to use the silliest method to solve it: cut this rectangular space just like a rectangular cake. The method although is a bit silly and old fashion, but it can cut the area quite efficiently. The first cut horizontally, to separate the first hall, lifted high to the end. The inner area, vertically with two cut; separate it into three natural vertical rectangle areas. The middle area leaves it as lifted high vertical space for visually and moving motion path integrating and merging. Two side each with one horizontal cut, to form a meeting room and office independently. The final cut leaves it for the backcourt office, for secretary and other reception, also data storage area, etc. Area naturally formed with neatness and simplicity. After dealing with this space, I find myself sometimes really appreciate this kind of neat and simple method to solve problem, without any prolixity. Mies Van De Rode said "Less is More", but honestly I don't really think it the same way. I rather agree with "Less is Good!", without any explanation and nonsense wording. Of course, it must according to the situation and the context back and forth to give conclusion, at least for a lamp and lanterns sales spacing, I definitely consider "Less, is definitely good".

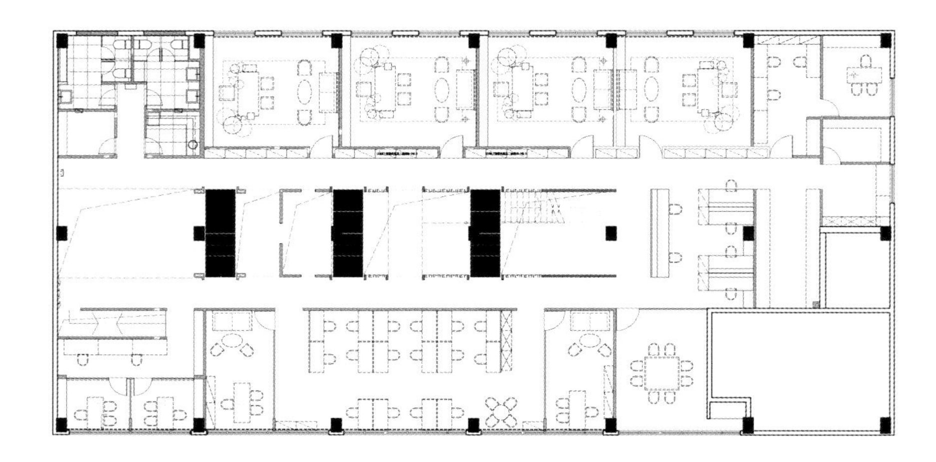

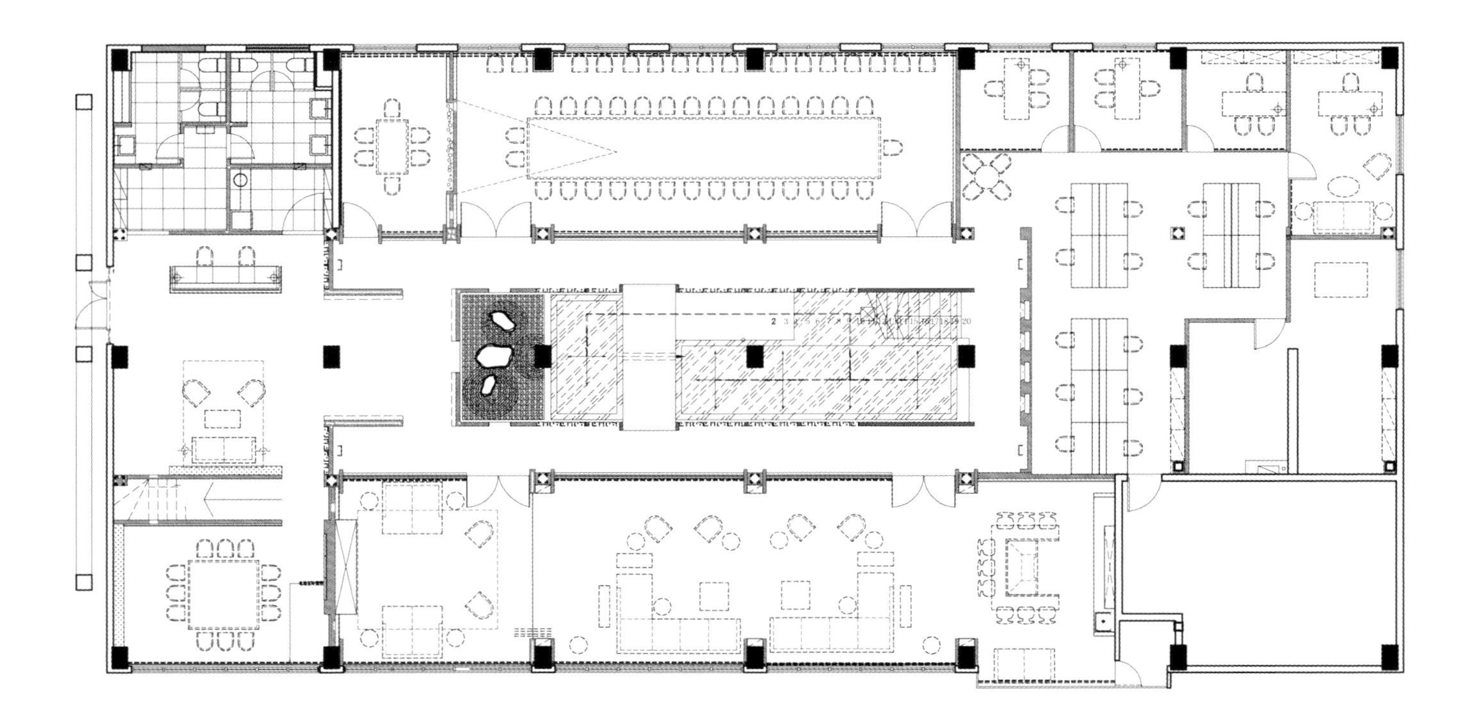

luminaire

Skin Peep

13

Design Company: Chris Briffa Architects

Project Location: St. James Cavalier, Valletta, MALTA

Area : 42 m^2.

Photographer: Chris Briffa

Skin Peep was born out of the idea of exhibiting the architectural works produced by Chris Briffa Architects in its first five years of existence along with the underlying process which brought them into being. The idea therefore came about to construct a temporary studio within the main hall at St. James Cavalier in Valletta to literally house this process and the people behind it, allowing visitors to peep in and watch the team at work.

Architecture starts when you carefully put two bricks together. There it begins. (Mies van der Rohe)

We like this famous quote, but it wasn't too influential on our choice of materials. We wanted to present the raw, often disliked, concrete brick in a new light. The brick provides holes for our audience to peep in, whilst their harsh skin provides protection to keep them out. The bricks will all be recycled back into the industry. The concealed timber supporting them is recycled formwork from one of our sites.

The idea of living and working in an art museum, under the gaze of strangers, might be regarded as exhibitionist behaviour. We tend to look at it as experimental motivation. Not so much for us (our inspiration comes from a dozen other influences, sometimes the most bizarre) but more for our viewers. We hope that our unorthodox workspace will inspire the mainstream enthusiast.

We do not think our work is minimal, or that it follows any particular style, but as our monolithic pavilion probably typifies, we tend to try making things look simple.

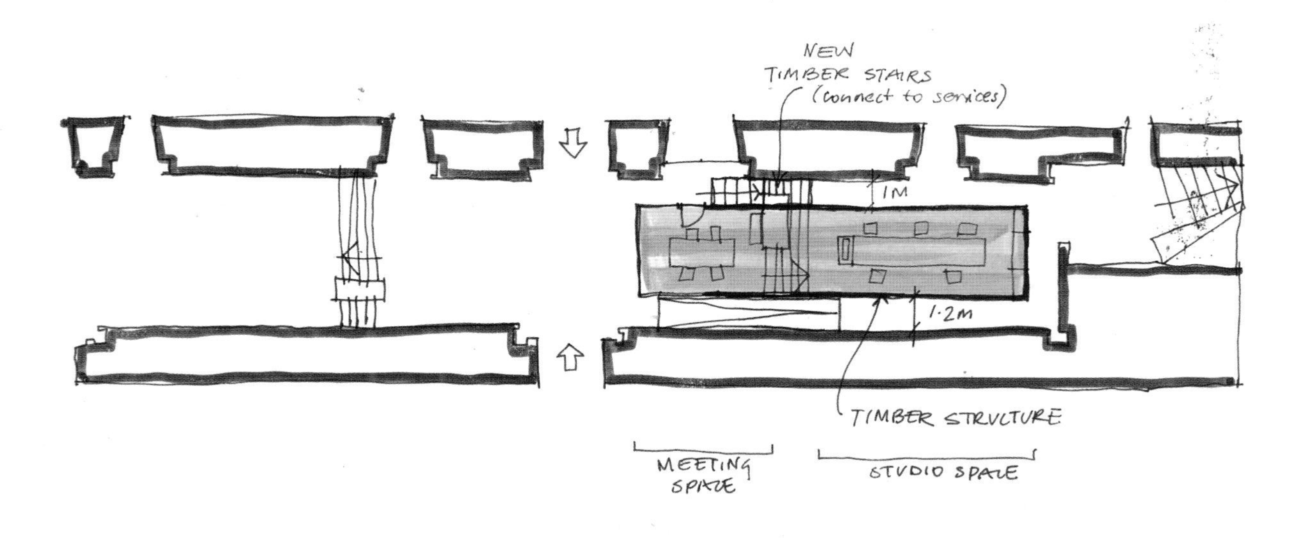
NEW
TIMBER STAIRS
(connect to services)
1M
1·2M
TIMBER STRUCTURE
MEETING
SPACE
STUDIO SPACE

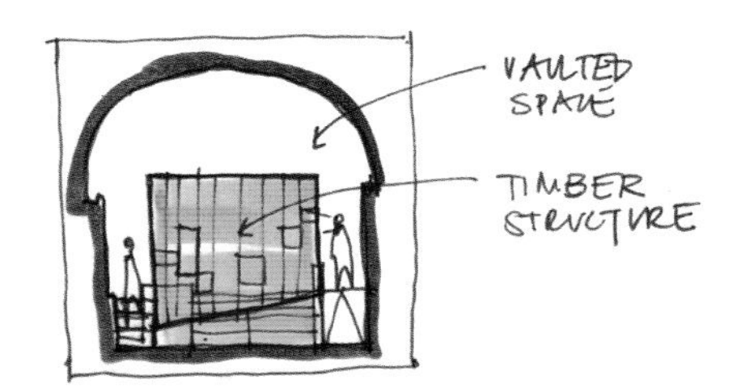
VAULTED
SPACE
TIMBER
STRUCTURE

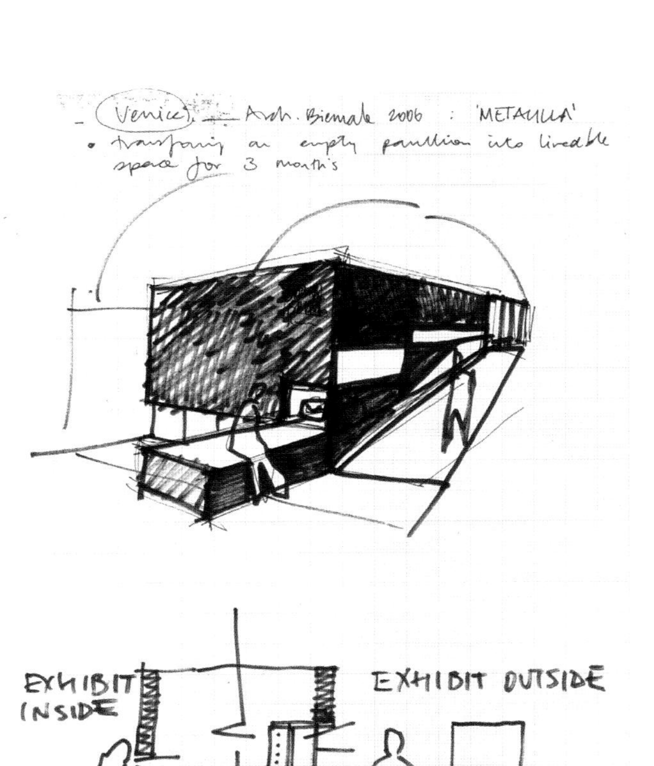

– Venice – Arch. Biennale 2006 : 'METALLICA'
• transforming an empty pavillion into liveable space for 3 month's
EXHIBIT INSIDE
EXHIBIT OUTSIDE
1.6M

Office Plajer & Franz Studio

14

Design Company: plajer & franz studio
Project Location: Berlin Germany
Area : 1000 m^2
Photographer: Diephotodesigner.de ken schluchtmann

The creative base of plajer & franz studio's 45 architects, interior and graphic designers is located in one of kreuzberg's largest industrial complexes dating back to 1898. their sensitive approach to the listed building clearly reflects the love of detail and sense of style they are so renowned for.

The striking mix of old and new is apparent as soon as you enter the light flooded reception area: existing ribbed slab ceilings and smooth suspended surfaces contrast each other as do the whitewashed stone walls and smooth plastered elements in warm mud colours, rough mastic asphalt flooring and customised furniture with sleek surfaces.

The monochrome colour scheme that pervades throughout the office is also clearly distinguishable: white and light coloured surfaces in combination with dark wood create a pleasant, neutral background for the individual colour accents of presentations, publications and unique collector's items. the use of indirect lighting and punctual accents of light further underline the light/dark theme.

The contoured reception counter and curved wall covered in alcantara form an inviting gesture. at the same time, functionality has not been neglected: situated behind the wall is a storage area for office materials, the effective backlight also serves as lighting for the storage.

The open loft-like workspaces are lined with extremely large window fronts, where individual workstations are docked on. airy white curtains allow for separation of specific work groups when necessary. sideboards along the windows offer ample storage and workspace for all creative tasks while tall industrial storage racks cover the opposing walls.

The two floors of the office are connected by minimalist filigree stairs reminiscent of a gangway. instead of being protected by a conventional railing, the floor opening is bordered with high boards and a stylised zen garden. dynamically shifted vertical light strips connect wall, floor and ceiling thus creating an eye-catcher at the end of the stairs.

The conference rooms are only separated from other work areas by frameless glass walls with specially designed door frames that seem to be floating in space. satined surfaces and curtains allow for more or less privacy. all the latest technologies from w-lan to beamer projection are concealed within the walls, ceilings and fixtures to let the clear new design and the existing architecture come to the fore. doors designed as acoustic panels are only discernible as vertical colour accents within the space, while individual wall sections open up to store documents and beverages. presentation boards can be placed on magnetic wall panels or narrow wooden shelves.

On the top floor, a curtain guided in a spiral rail enables a flexible use of the space. depending on its position, a more or less enclosed semitransparent but cosy conference space is created. the fabric and carpet give the room pleasant subdued acoustics.

For smaller client meetings, the lounge in the executive office is a perfectly peaceful oasis. the couch, carpet, warm colour scheme and decorative elements create a homey environment. the iridescent blue fish of the built-in freshwater aquarium also have a soothing effect. all media technology is out of view, concealed behind the brass wall panels. those holding the aquarium are sound insulated and can be opened completely for maintenance.

The heart of the studio is the open kitchen. this is where the architects' philosophy comes to life. whether it's at the bar, on the couch or at the big table, communication and exchange of ideas is the key. during meetings, for breakfast or while cooking a meal together, the professional and social exchange here takes place in a relaxed atmosphere. on special occasions, integrated speakers and media technology can transform the space into a club or home cinema. with floor to ceiling sliding doors that disappear completely within the walls, the space can be shut off from the rest of the office.

Annexed to the lounge is the studio's "treasure chamber", where the latest innovative materials and samples of the newest technological developments fill floor to ceiling storage racks.

The combination of function and aesthetics as well as the contrast between the industrial character of the building and the sophisticated new interior design, carry the signature of the berlin architects: it's always the rough with the smooth and the hot with the cold, but never the lukewarm.

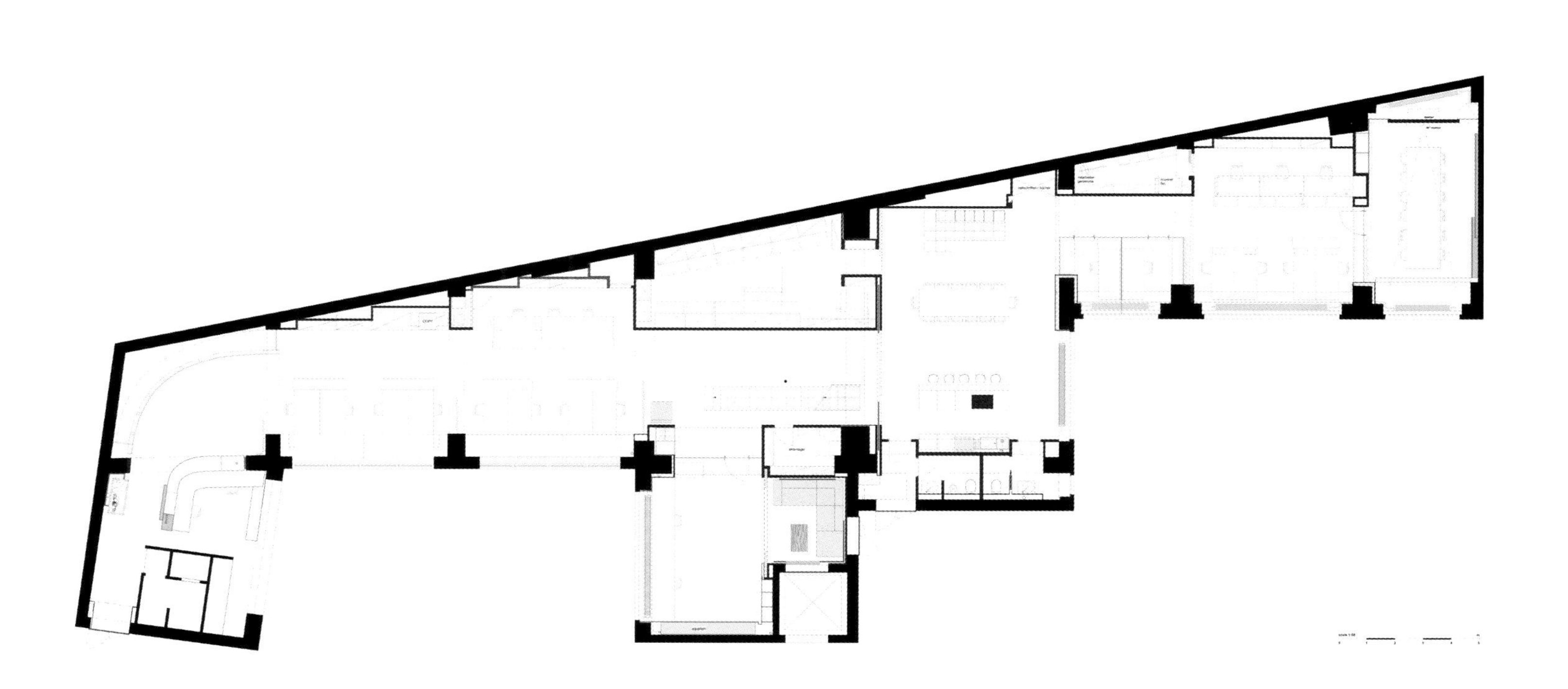

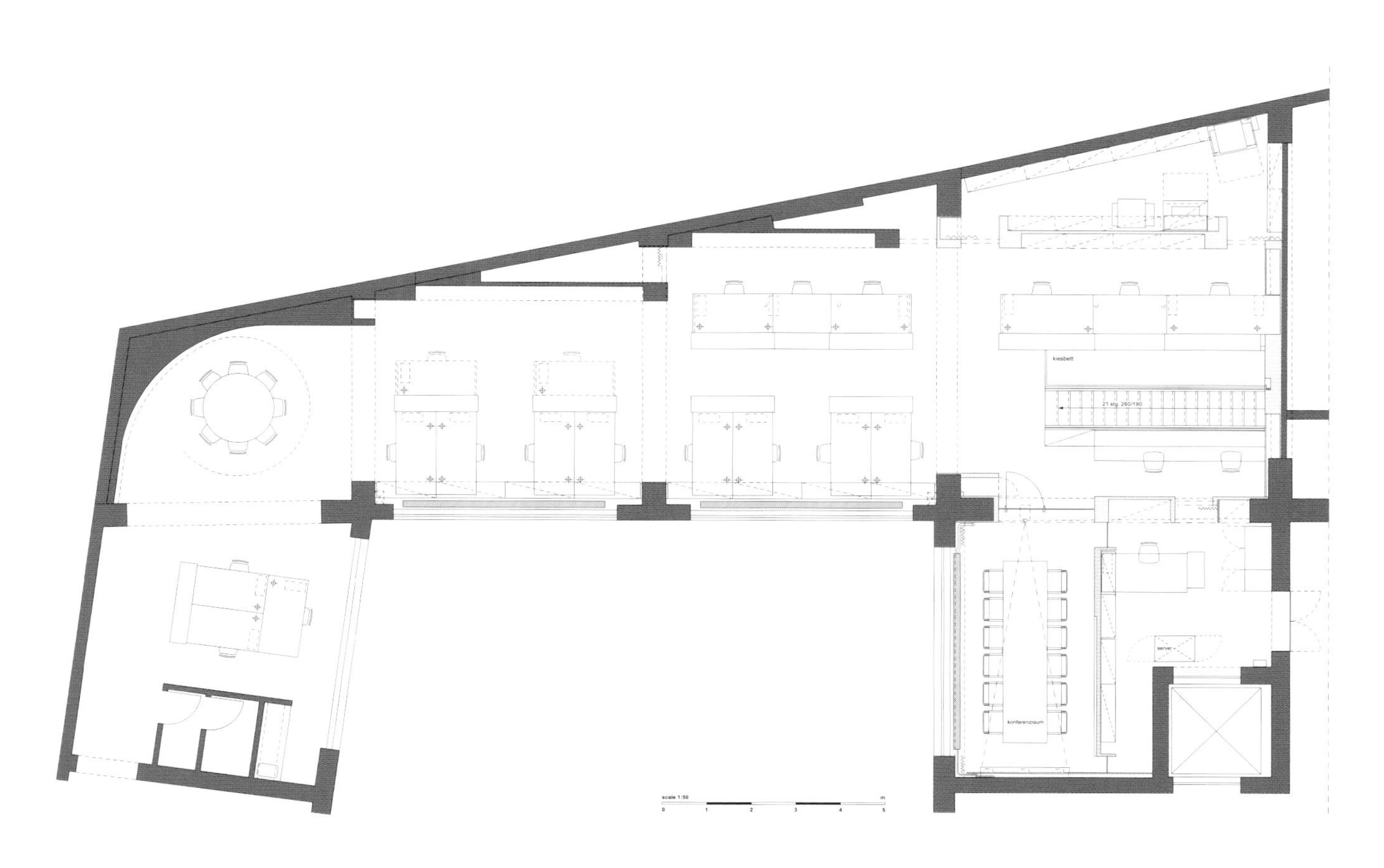
kiesbett
21 stg. 260/190
server
konferenzraum
scale 1:50
m
0
1
2
3
4
5

Leo Burnett Office

15

Design Company: Duangrit Bunnag Architect Limited – DBALP

Project Location: Pathumwan, Bangkok

Area : 370 m^2

Photographer:Mr. Wison Tungthunya

Leo Burnett is a international advertisement agency with a unique attitude. They always believe in the simplicity towards the unexpected creativity. Those become the main idea of their Bangkok office renovation in 2002.

The 4,000 square metre space was originally used as a stock market exchange with loft like space and high ceiling. The idea is to play with clean geometry and blunt contrast of materiality. The space also suggests the high ceiling analogy of 'creative factory' and the design is following as such.

The office area is based on the open planning concept, which is a contradictory idea of local use. But after a few years in operation, the concept has been proven on its originality as well as it practicality. Together with the modular grid planning concept, it allows the space to be flexibly adjustable over time.

RMS CALIFORNIA

16

Design Company: Resolution 4 Architecture

Project Location: Newark, CA

Area: 9290 m^2

Photographer: Bernard Andre

Directed by the vision of Hemant Shah, CEO and founder of Risk Management Solutions, RES4 was requested to implement an environmental transformation into the design of RMS's new U.S. Headquarters, similar to the work completed by RES4 at their Hackensack, NJ office. Despite the ongoing economic crisis, the company's growing success was taken in high consideration during the master planning over the next 5-10 years for the potential 440 employees. By creating smaller, more efficient private workspaces, it allows for larger communal and conferencing spaces in results of a creatively competitive and inspiring ambiance. In response to RMS's technological advances, a 24- monitor media wall is a primary focal point, in efforts to continuously project a library of current catastrophic bullets and targets, relating to the company's purpose of providing products, services, and expertise for the quantification and management of catastrophe risk. With centralizing the majority of enclosed spaces and glazing all peripheral faces, visual transparency from all internal points to an exterior band of fenestrations and natural light is achieved. Infiltration of natural light reduces the need for artificial light in the open workstations surrounding the exterior shell. RMS's global awareness is reflected through the materials & finish choices as well as the low-energy/ consumption of appliances & plumbing fixtures.

EXIT

EXIT

UXUS Office

17

Design Company: UXUS

Project Location: Amsterdam, Netherlands

Area: 300 m^2

Photographer : UXUS

The theme for the new UXUS office's décor is inspired by the rich visual world of Fables. Both the fairytale location of its Art-Nouveau/Neo-Gothic office tower at the crossing of two major canals in central Amsterdam, and the design agency's penchant for the mysterious and poetic atmosphere of old and new world Fables, influenced the look and feel of a space where imagination is free to roam.

As Fables bring together narrative and memorable insights into the everyday, UXUS has created an environment to stimulate the imagination and foster powerful, unforgettable stories that color our everyday life.

About UXUS

Founded in Amsterdam in 2003, UXUS is an independent award wining design consultancy specializing in strategic design solutions for Retail, Communication, Hospitality, Architecture and Interiors.

UXUS creates "Brand Poetry", fusing together art and design, and creating new brand experiences for its clients worldwide. We define "Brand Poetry" as an artistic solution for commercial needs.

Artistic solutions target emotions; emotions connect people in a meaningful way.

Design gives function, art gives meaning, poetry expresses the essence.

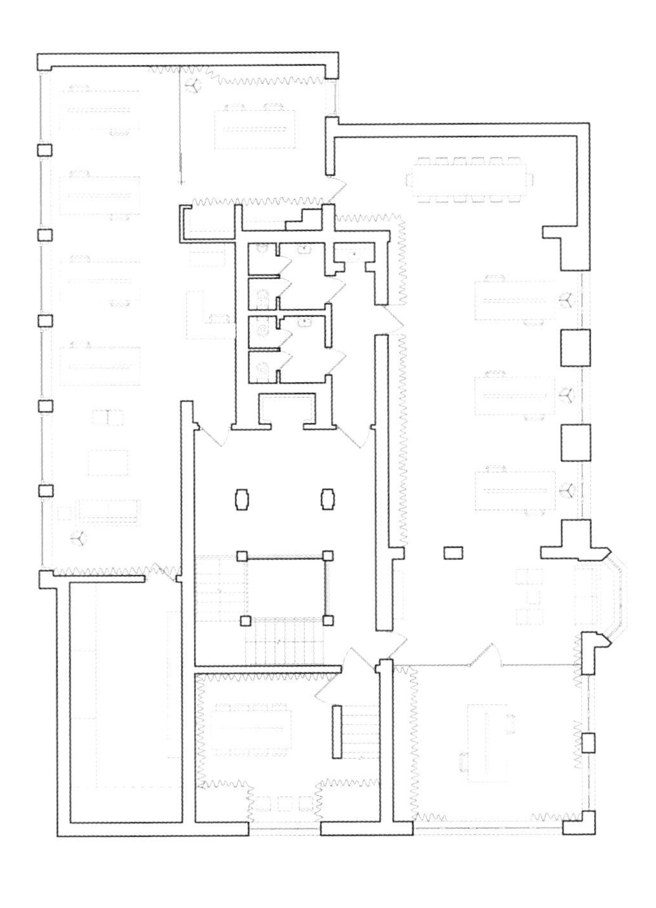

Brunete's Office

18_

Design Company: Brunete Fraccaroli Interior Architecture

Project Location: São Paulo

Area: 210 m^2

Photographer: Tucá Reines

Brunete Fraccaroli´s proposal to developing her office, was based on design a place to work, that was joyful and confortable. The colors became from Solutia´s films, associated to the utilization of gass, these are the strengths explored.

The place´s audacity is noted since from the entrance, with a spread hall. The colors plays between the green, blue and pink, and produces a funny air to the office, making it playful and dynamic. Added to the transparency of the glass, creates a space contemporaneous and clean, while the mirrors give a sophisticated touch to the room.

The architect privileged the ideia of turn the office functional and pleasant to work. Therefore, the silver cabinets made in polyurethane were made by Madeira Viva, are full of refinement. The entrance doors have its open by biometrics, which facilitates the movement, installed by Cynthron. The lightning, from La Lampe, values the colors from the place. The audio e video´s project integrated all the office and was executed by Josias Studio.

These characteristics are the translation from her way of living, happy and sophisticated, always searching for perfection in all of the details, producing a place which reflects the architect Brunete Fraccaroli in her single way.

PRISMACOLOR
PRISMACOLOR

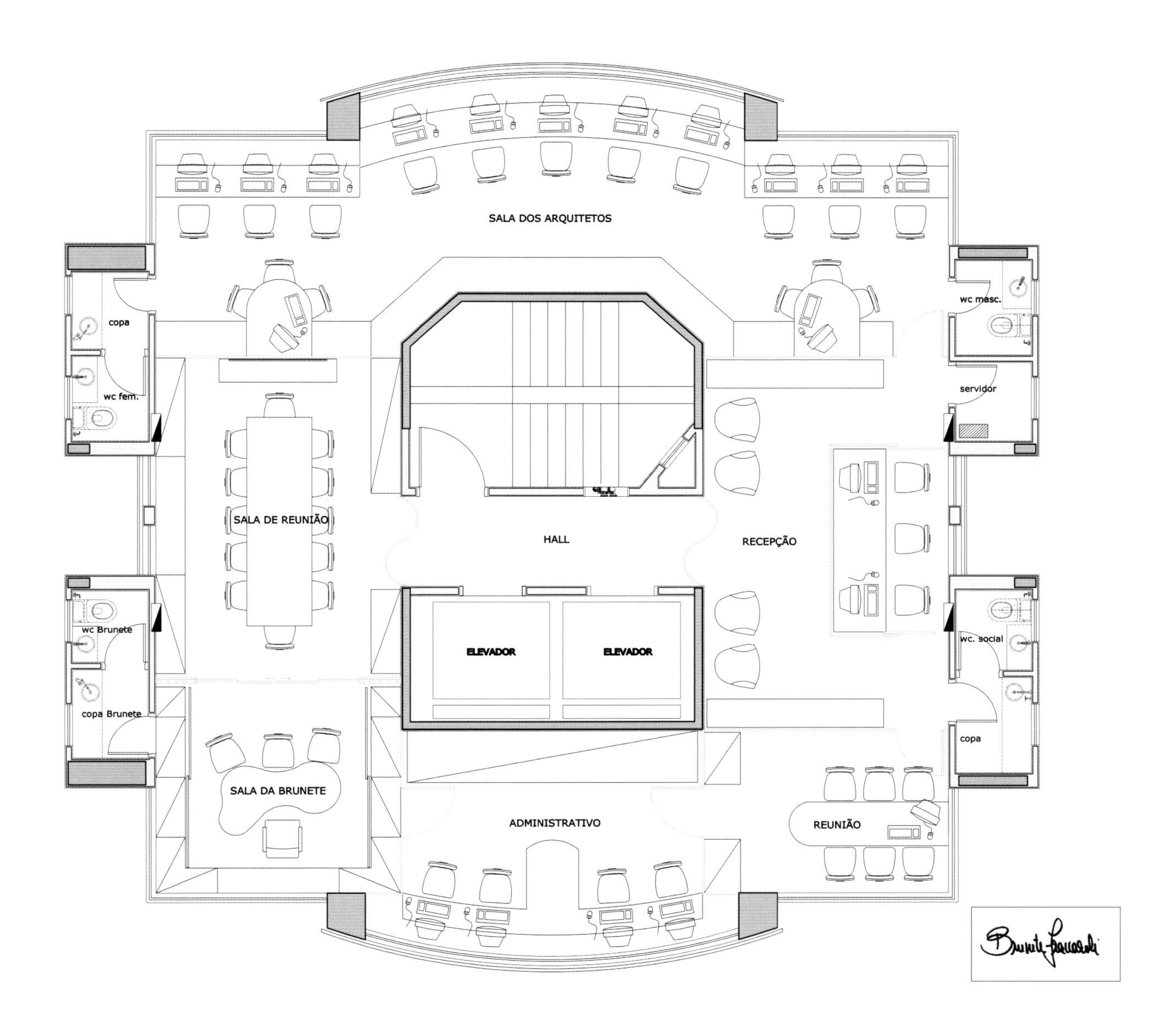
SALA DOS ARQUITETOS
copa
wc fem.
wc masc.
servidor
SALA DE REUNIÃO
HALL
RECEPÇÃO
ELEVADOR
ELEVADOR
wc Brunete
copa Brunete
wc. social
copa
SALA DA BRUNETE
ADMINISTRATIVO
REUNIÃO

Design Army Headquarters

19

Design Company: Studio Twenty Seven Architecture
Project Location: Washington
Area: 483.08 m^2
Photographer: Anice Hoachlander, Hoachlander Davis Photography, www.hdphoto.com Erik Johnson Photography, Inc.

In 2006, Design Army, an accolade laden Washington DC graphic design firm looked for a new space to locate its growing business. After a concentrated search, Design Army was able to purchase an abandoned and dilapidated building in the City's NOMA (North of Massachusetts Avenue) District. Destroyed in the riots of 1968, the NOMA District had laid dormant for over 30 years before being identified as a focus redevelopment zone in the City's master plan.

Programmatically, Design Army needed to gut and expand the structure from a two to a three and a half story building. The eight person firm needed two and half floors of studio space as well as a street-front retail space they could lease to supplement the operating costs of the new building. The architects were aware that the new zoning overlay about to be adopted for the neighborhood would make this impossible by restricting the building to three floors. The architects were able to work with the planning commission and zoning board to obtain a variance to allow a mezzanine roof deck and kitchen area for Design Army to use and make sure the emerging neighborhood did not lose this new development. Design Army represented a vibrant new source of activity and anchor redevelopment property for the block.

In 2008, Design Army took possession of a luminous work and meeting space that reinforces their clients' belief that they have hired a graphic design firm that provides "designs that resonate with intelligence, relevance and professionalism."

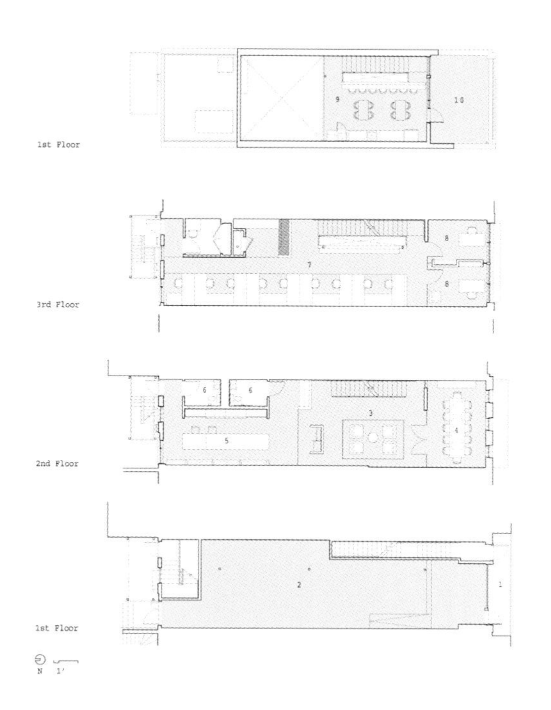

1_Street
2_Sublet Space
3_Reception
4_Conference
5_Copy/Print/Cut
6_Bath
7_Studio
8_Office
9_Kitchen
10_Deck

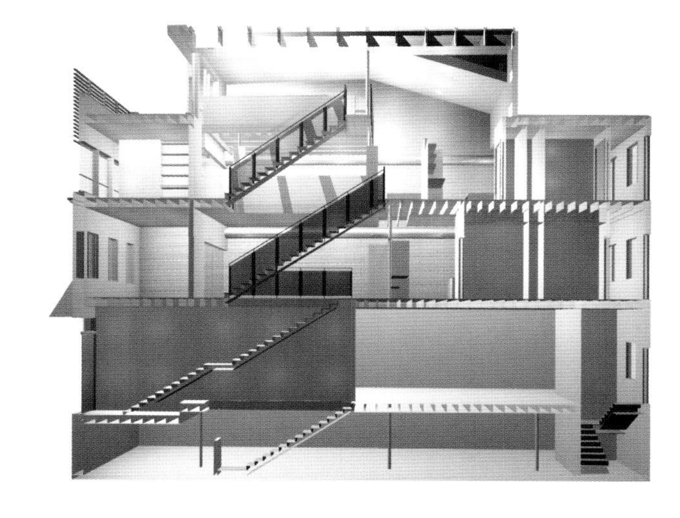

Bogart Lingerie

Design Company: UdA

Project Location: Hong Kong

Area : 2000 m^2

A leading South-East Asia third-party manufacturer of underwear wanted the new corporate headquarters to be made on the twentieth floor of a new office tower block overlooking Hong Kong bay. The plan takes up an overall 2,000 sqm surface; it includes an 800 sqm reception area giving access to board, meeting and show rooms in the nature of enclosed rooms, as well as a 1,200 open space where a design development department is hosted. A shared café area links these two portions of the building and contributes to the overall unity effect.

The starting concept puts a new focus on the way the big corporations operating in South-East Asia but at the same time possessing a global scope want their headquarters to be seen by the public eye, beyond traditional, stereotyped appearance. A design pattern deconstructing the usual myth was sought for, which would as much let the company's double identity – the European and the Asian ones – as the double-sided nature of the spaces emerge clearly: the male universe of the business centre embodied by the building and its equipment (well represented by the existing aluminium false ceiling that cannot be removed) and the company's mission addressing the brassiere, or more commonly bra, a garment that has become an icon of modern refined sophisticated womanhood.

Thus the inner atmospheres, the materials used as well as the furniture all reveal lightness and brightness much in the North-European style, especially in the white surfaces and the light-coloured wooden floor.

On this backdrop one can perceive subtle, sometimes a bit elusive suggestions, just as delicate but also complex as the company's products are: they peer through window panes that remind the onlooker of white laces, or cabinets and sample archives coated with printed patterns laminated cloth, or furniture that is at once light and remarkably imposing, made in Italy after UdA original design using linoleum surfaces and multilayer structures. The whole furniture range, from the reception desk to the cafeteria, to the boardrooms, reveals its constructive nature, a simple and at the same time precise aesthetics of surfaces and assembling methods – exactly like the feminine spirit in the world.

BOGART LINGERIE LTD

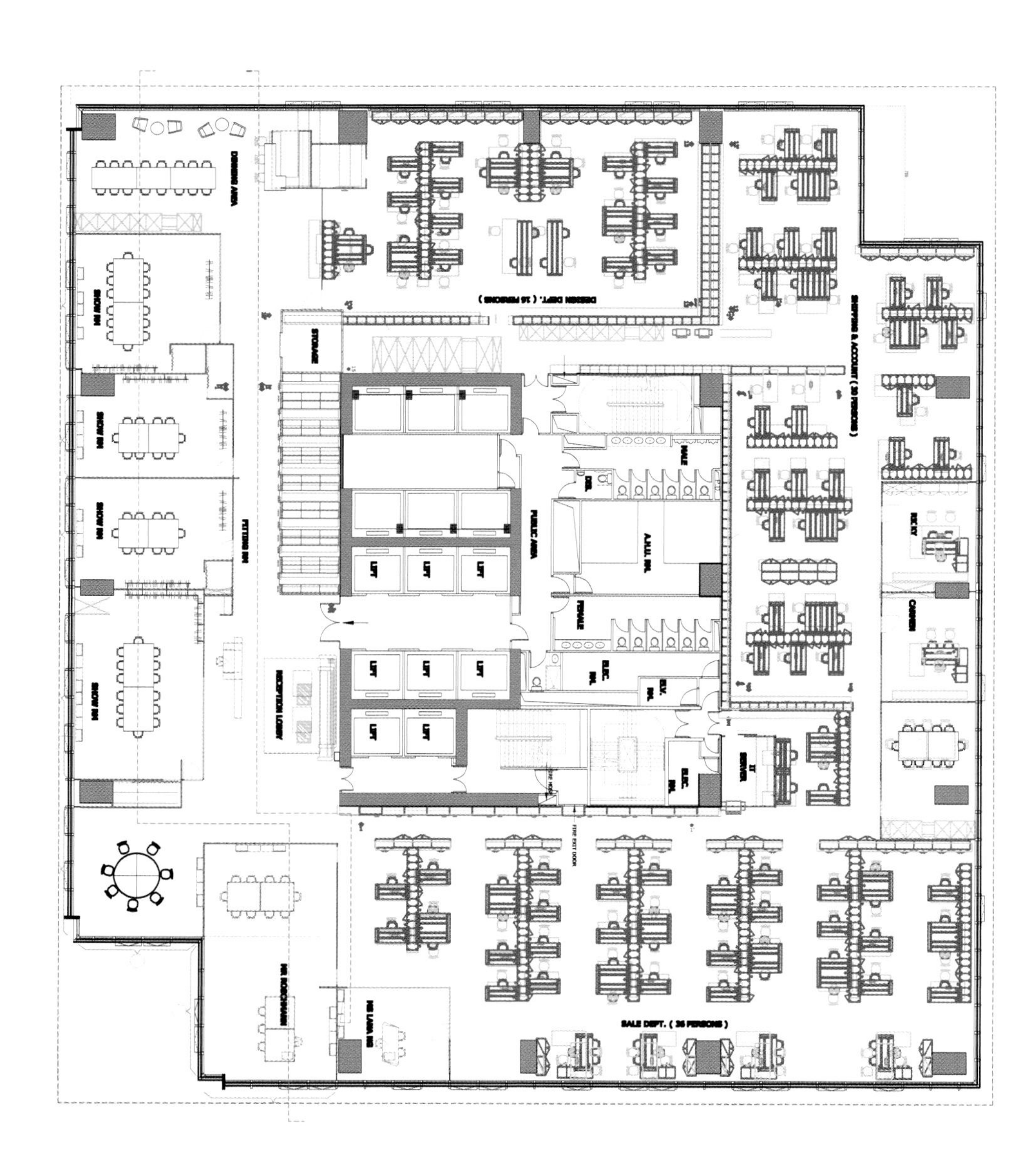

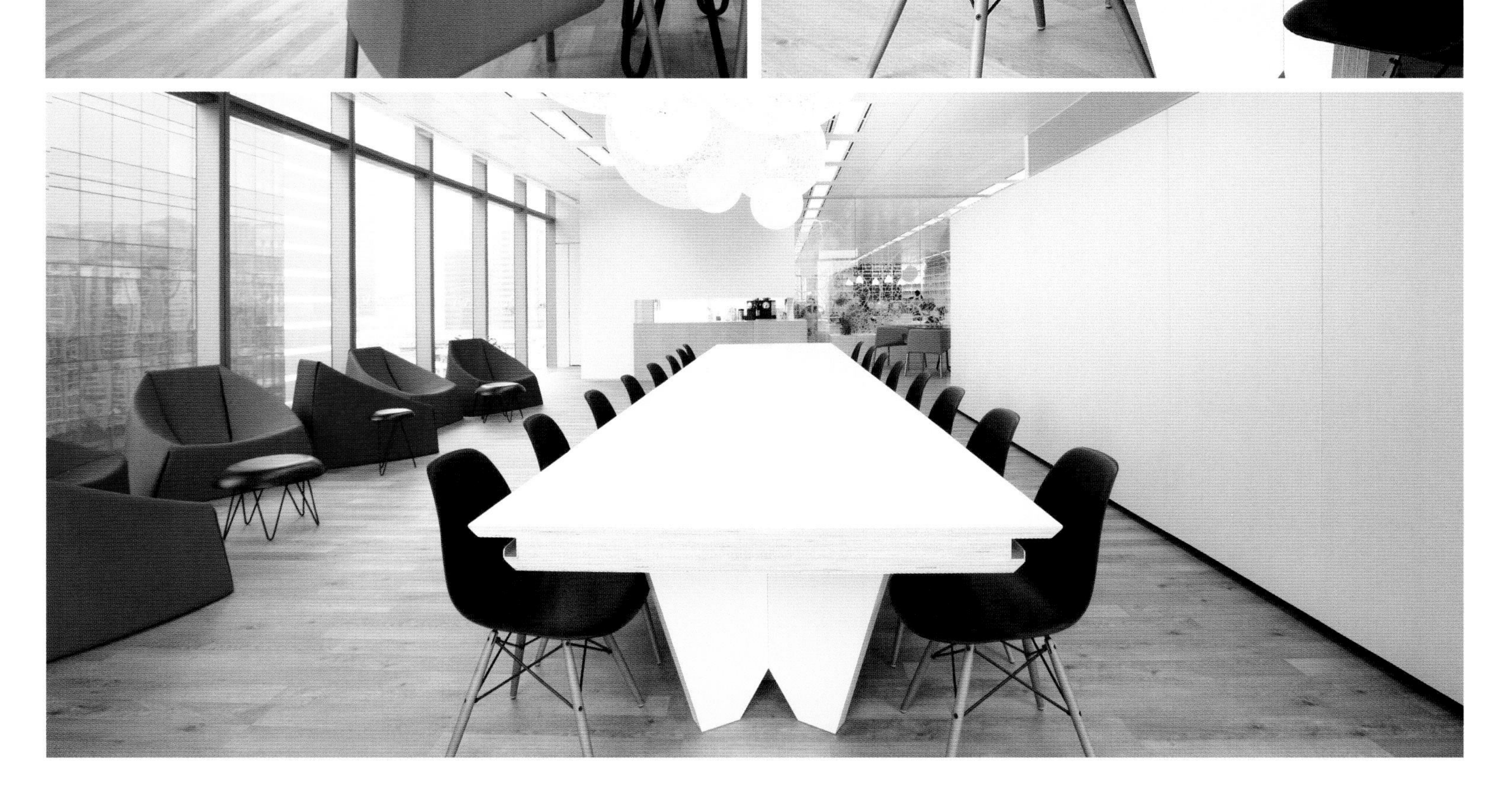

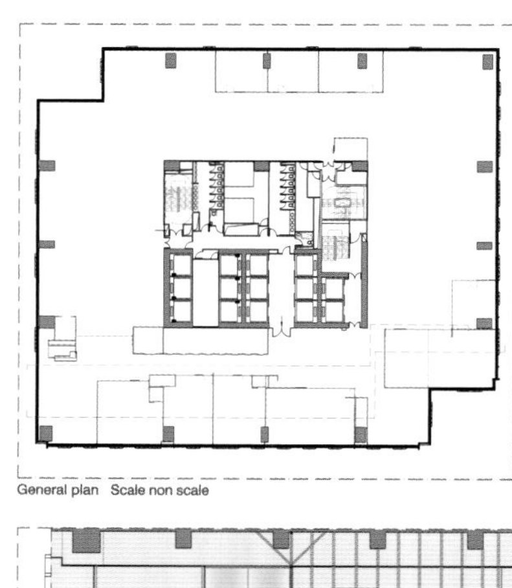

General plan Scale non scale

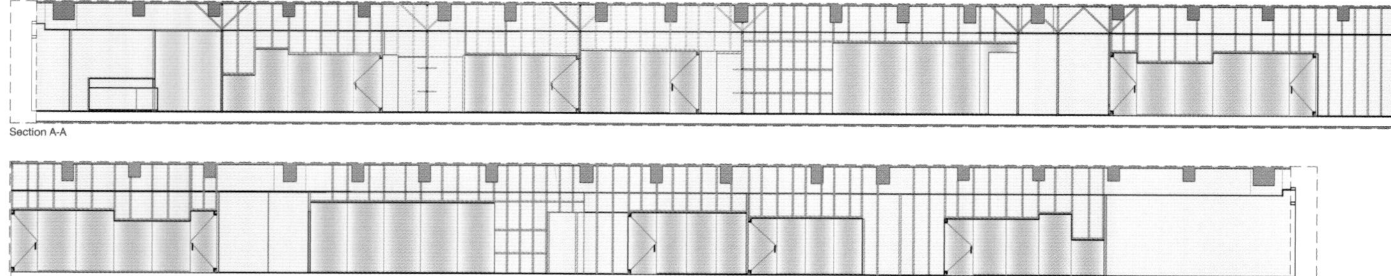

Section A-A

Section B-B

RI Offices

21

Design Company: arquitectura x

Project Location: Quito, Ecuador

Area : 82 m^2

Photographer:sebastian crespo

this small intervention for a young publicity agency is based on the need for flexible adaptable and expandable space for their multiple tasks and changing needs, within a very low budget, since a large investment had to be made on hardware, software and specialized photographic, film and editing equipment. parallel to cost reduction the scheme had to provide a strong and fresh image to the agency as this was their first formal office; our response was to remove all the standard anodyne finishes and leave all services and the concrete and steel structure bare, obtaining a flexible raw container where we could insert pieces of furniture or containers of activities that could be modified as needed in time. these furnishings plug into the modified exposed services and are made of maple veneer plywood and floating floor used for all surfaces, making them cheap and easy to install or modify.

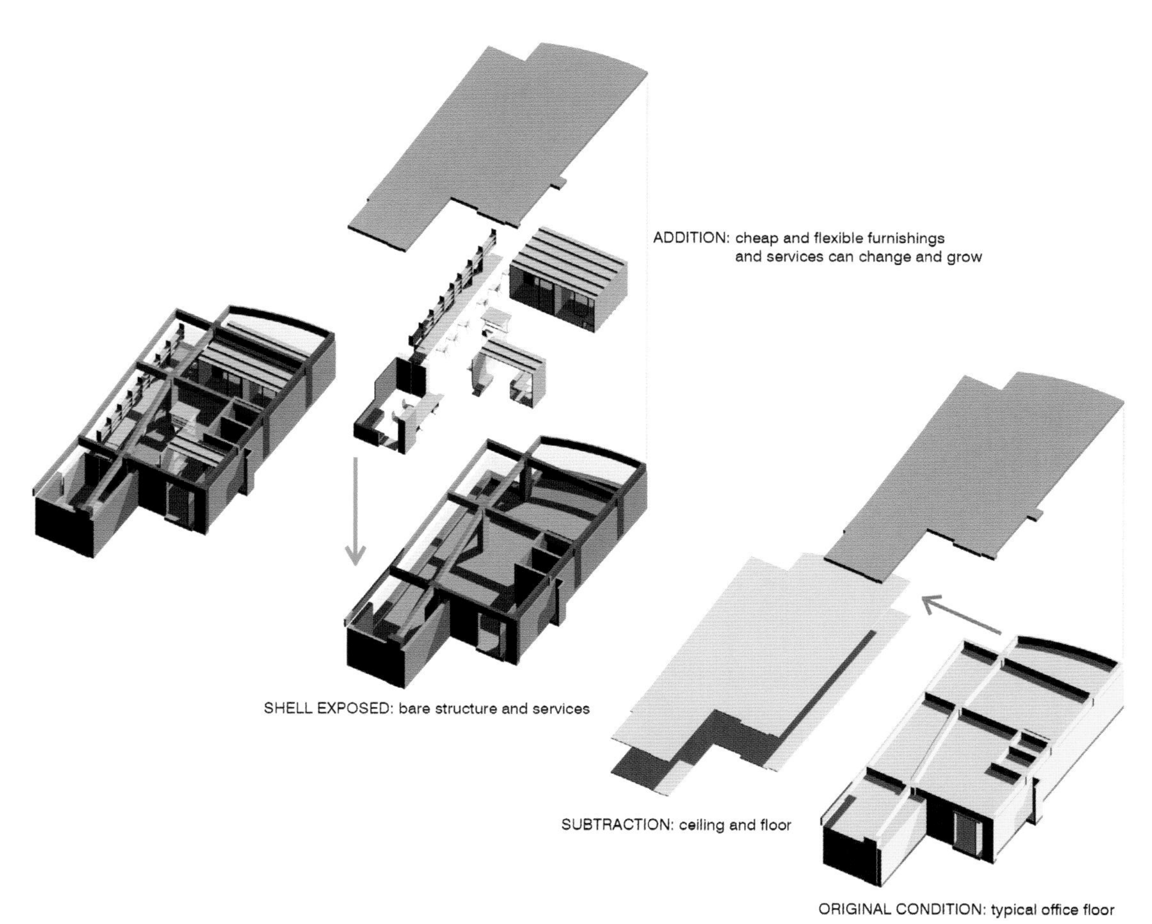
ADDITION: cheap and flexible furnishings
and services can change and grow
SHELL EXPOSED: bare structure and services
SUBTRACTION: ceiling and floor
ORIGINAL CONDITION: typical office floor

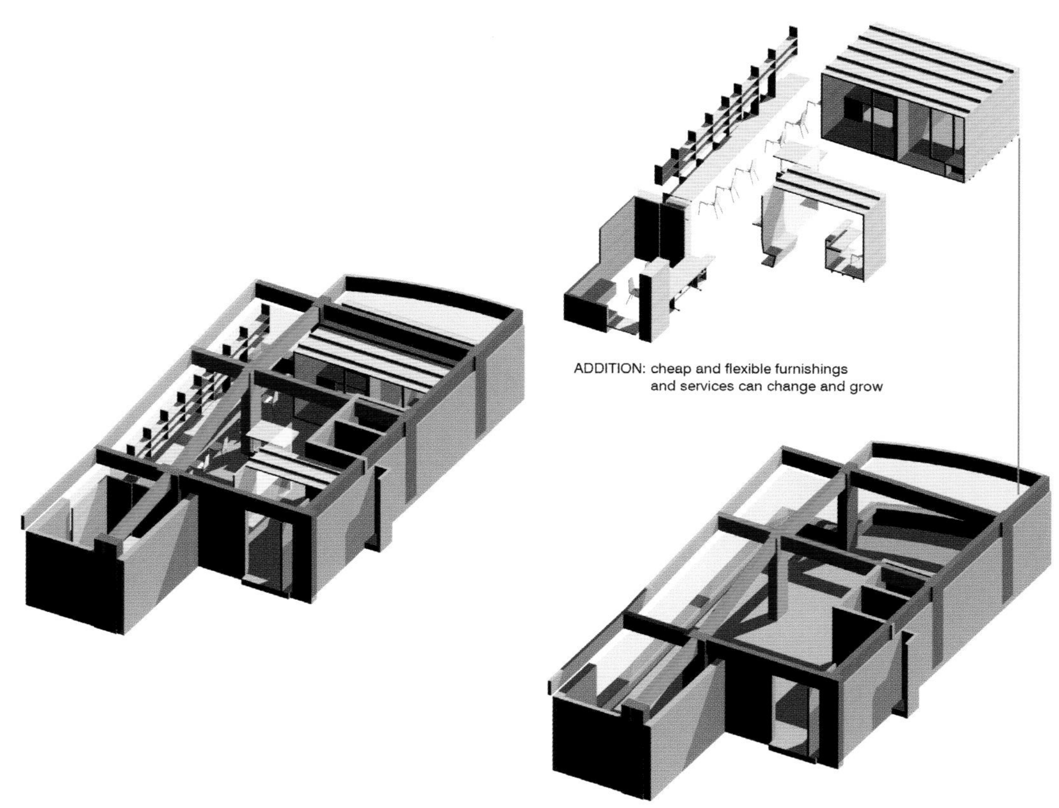
ADDITION: cheap and flexible furnishings
and services can change and grow
SHELL EXPOSED: bare structure and services

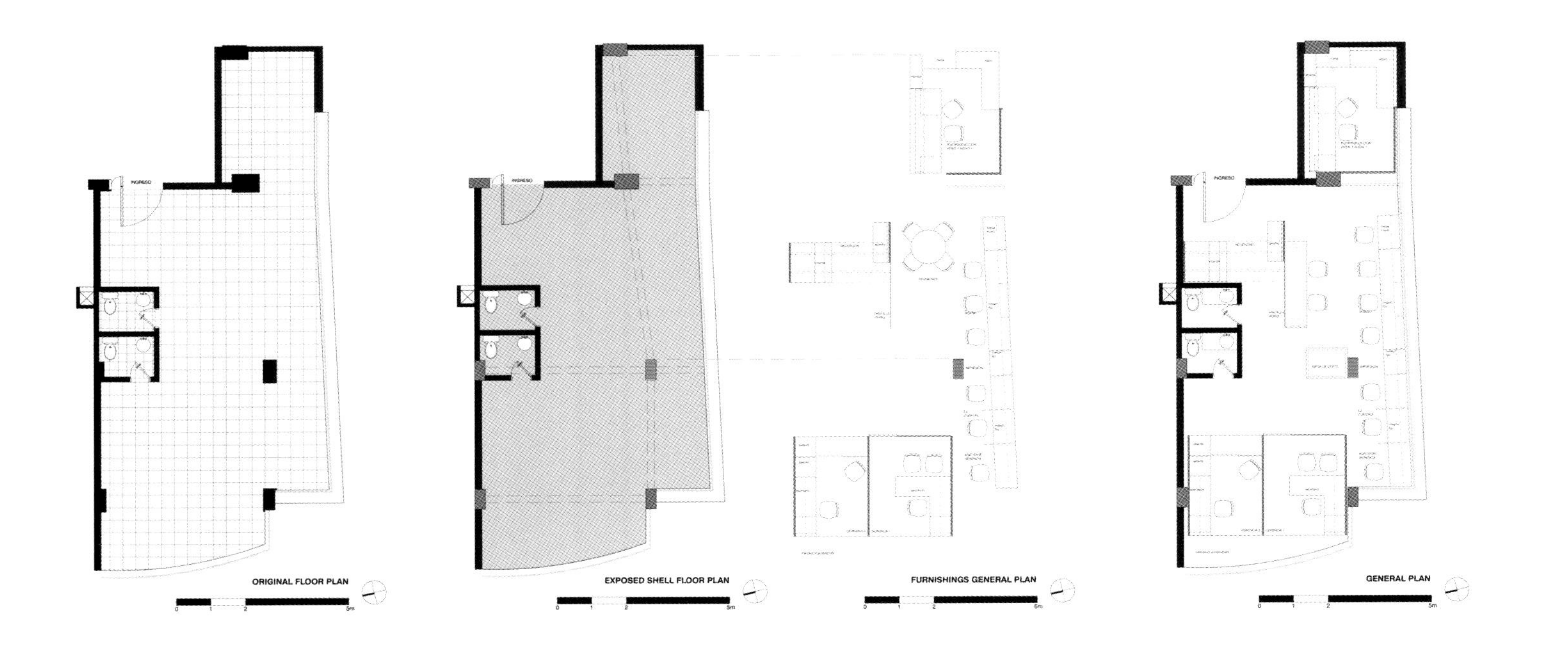
ORIGINAL FLOOR PLAN
EXPOSED SHELL FLOOR PLAN
FURNISHINGS GENERAL PLAN
GENERAL PLAN

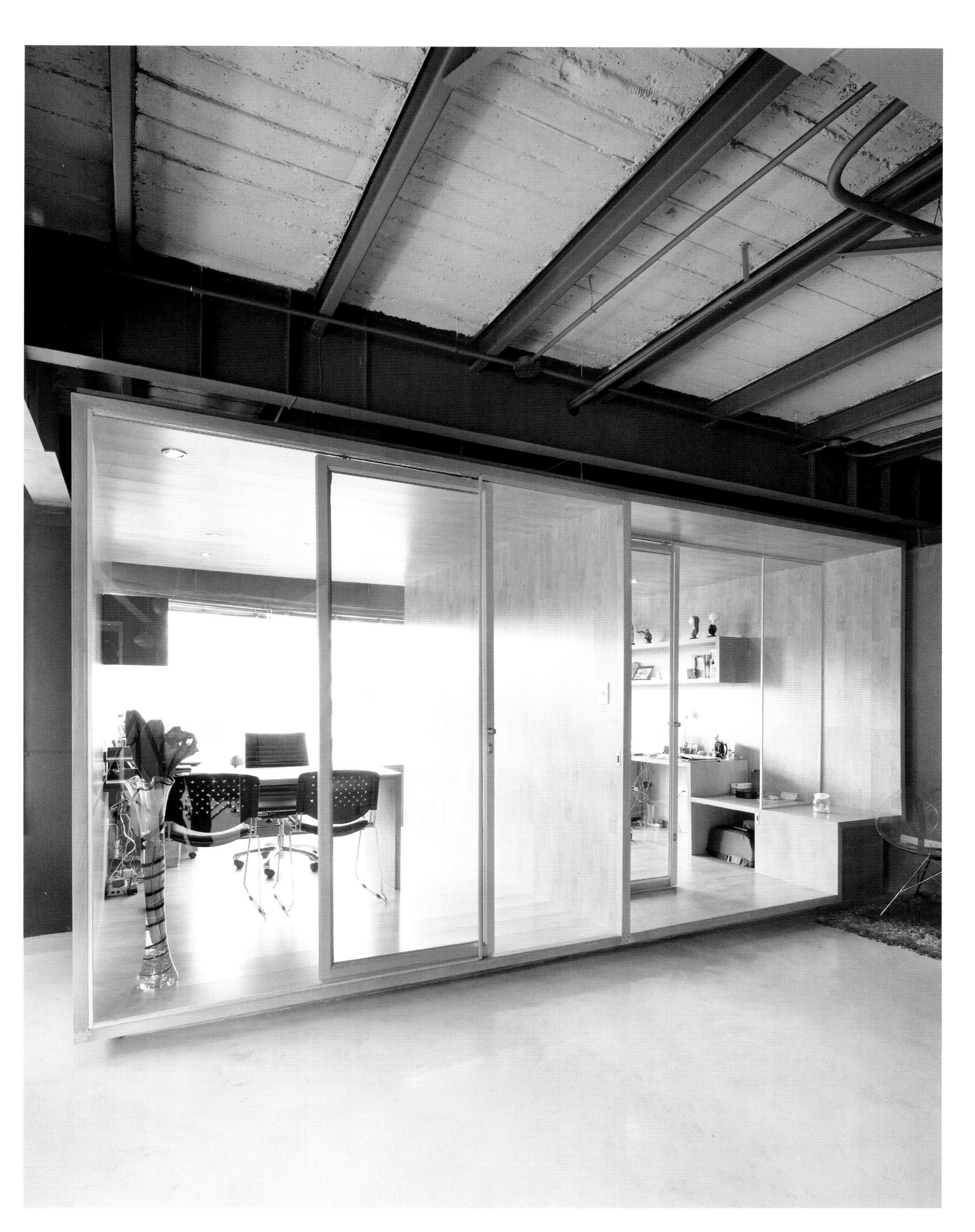

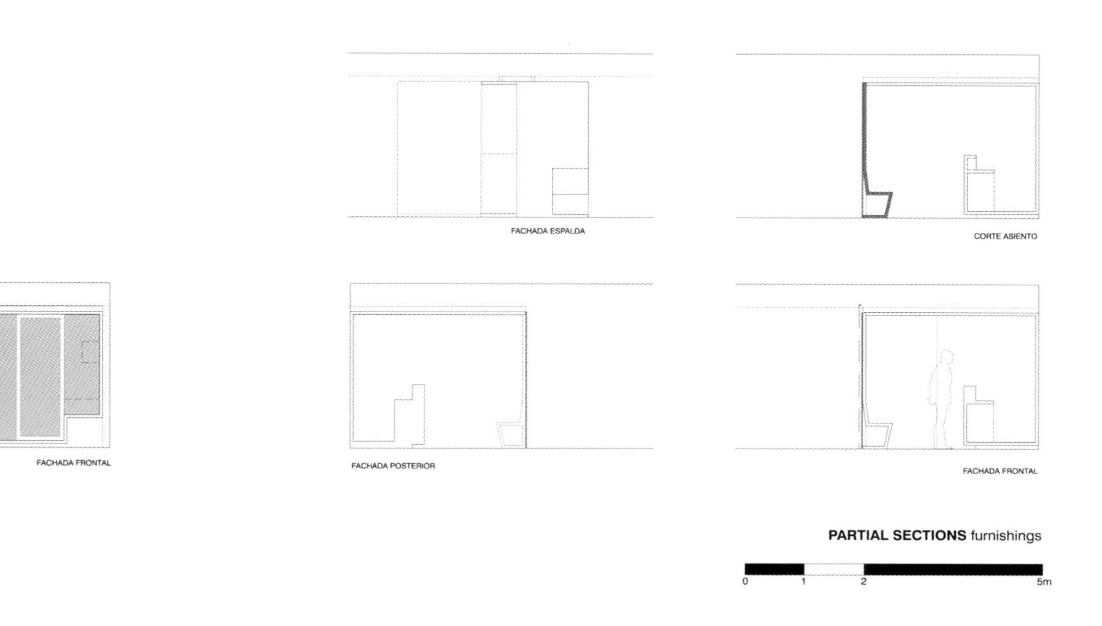

PARTIAL SECTIONS furnishings

Acbc office

22

Design Company: Pascal Arquitectos. Carlos y Gerard Pascal

Project Location: México

Area : 512 m^2

Photographer:Jaime Navarro

This office project was designed for a shopping center developer. The initial concept aimed to create a very functional and contemporary headquarters. The design intention was to build, through the language and the atmosphere, architecture completely different from the traditional, institutional approaches, and based on the premise of the workplace as a second home where people seem to spend most of their time.

A mix of natural materials such as marble and wood, in contrast with colored tempered glass, stainless steel and iron, created balanced color and textures compositions and at the same time offers a modern and a warm ambience.

At the elevators entrance hall, the difference between the white calacata marble and the black steel plates suggests a contradictory effect. This access leads to the reception foyer changes to a palette with elegant and warm tones and sets the atmosphere through its lighting. In this area: rosso levanto marble, papagayo wood wall panelling and a black crystal glass with a floating LCD screen that temporarily appears and disappears.

The reception foyer is the access to a hallway that communicates the rest of the sections of this complex with different level surfaces and materials that hide recessed lighting.

The operational section is a large hall with working stations. In this space, a mural decoration, alluding to the company's business, was placed along with an indoor carpet for acoustic damping. This section also displays crystal partitions that allow sunlight through the private rooms.

Located at the end of the hallway the secretarial reception of the executive area, featuring geometrical engraved crystal doors, a Jose Villalobos painting and Piranessi engravings. The executive also compromises two executive offices, a boardroom and a lounge room, all with modern furniture, wood wall panelling and ceilings in combination with plaster ceilings by grooves that hide air conditioning injections. Velvet draperies complement the atmosphere with a residential and elegant touch.

In the boardroom, same materials and ceilings are used, these last ones with direct and recessed lighting through onyx plates.

Finally, it is worth to mention that the complex also has an unusual area: a lounge room with a bar that offres the opportunity to have pleasant and informal meetings around an iron and arabescato marble bar.

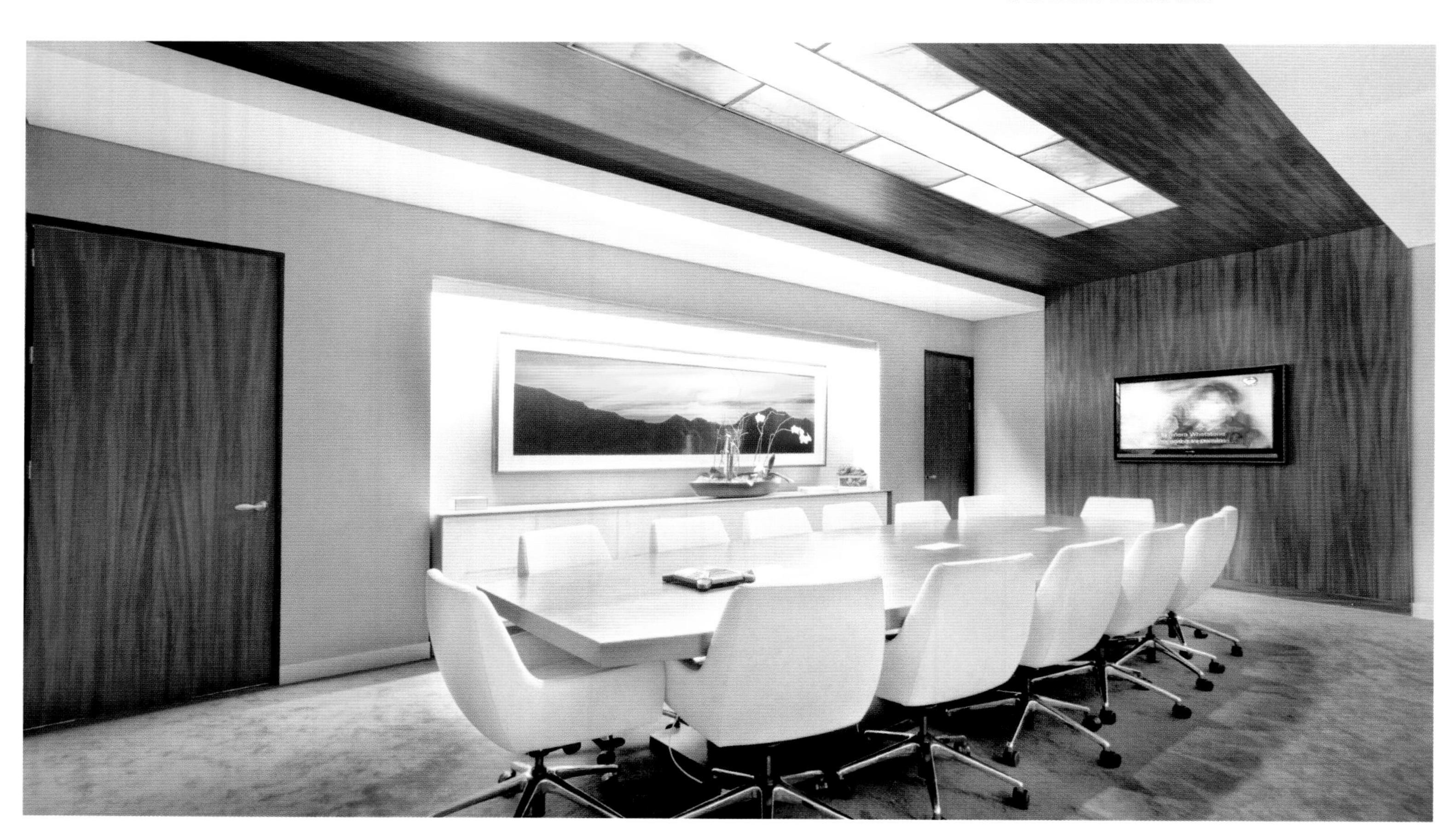

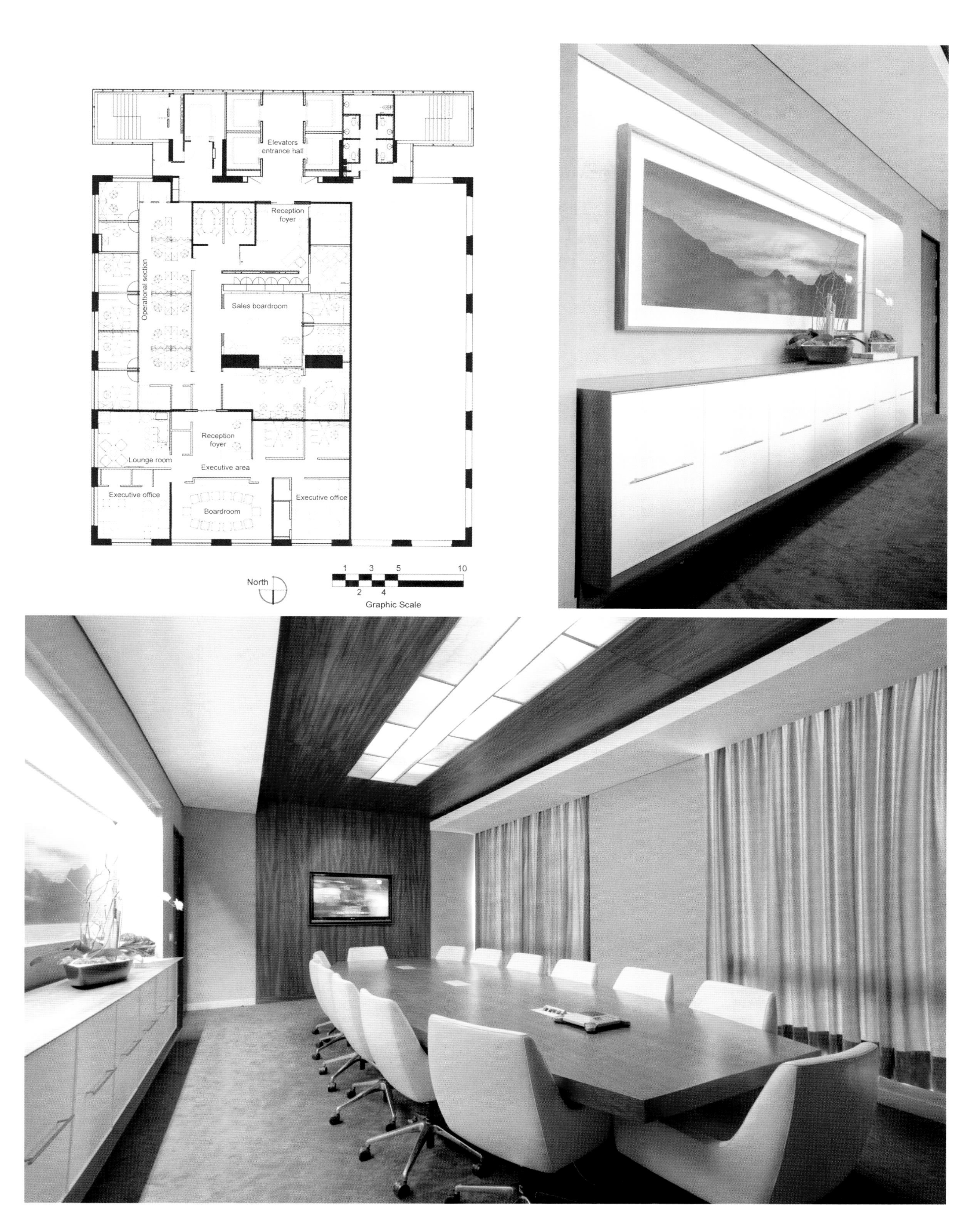
Elevators entrance hall
Reception foyer
Operational section
Sales boardroom
Reception foyer
Lounge room
Executive area
Executive office
Boardroom
Executive office
North
1 3 5 10
2 4
Graphic Scale

Bloomberg

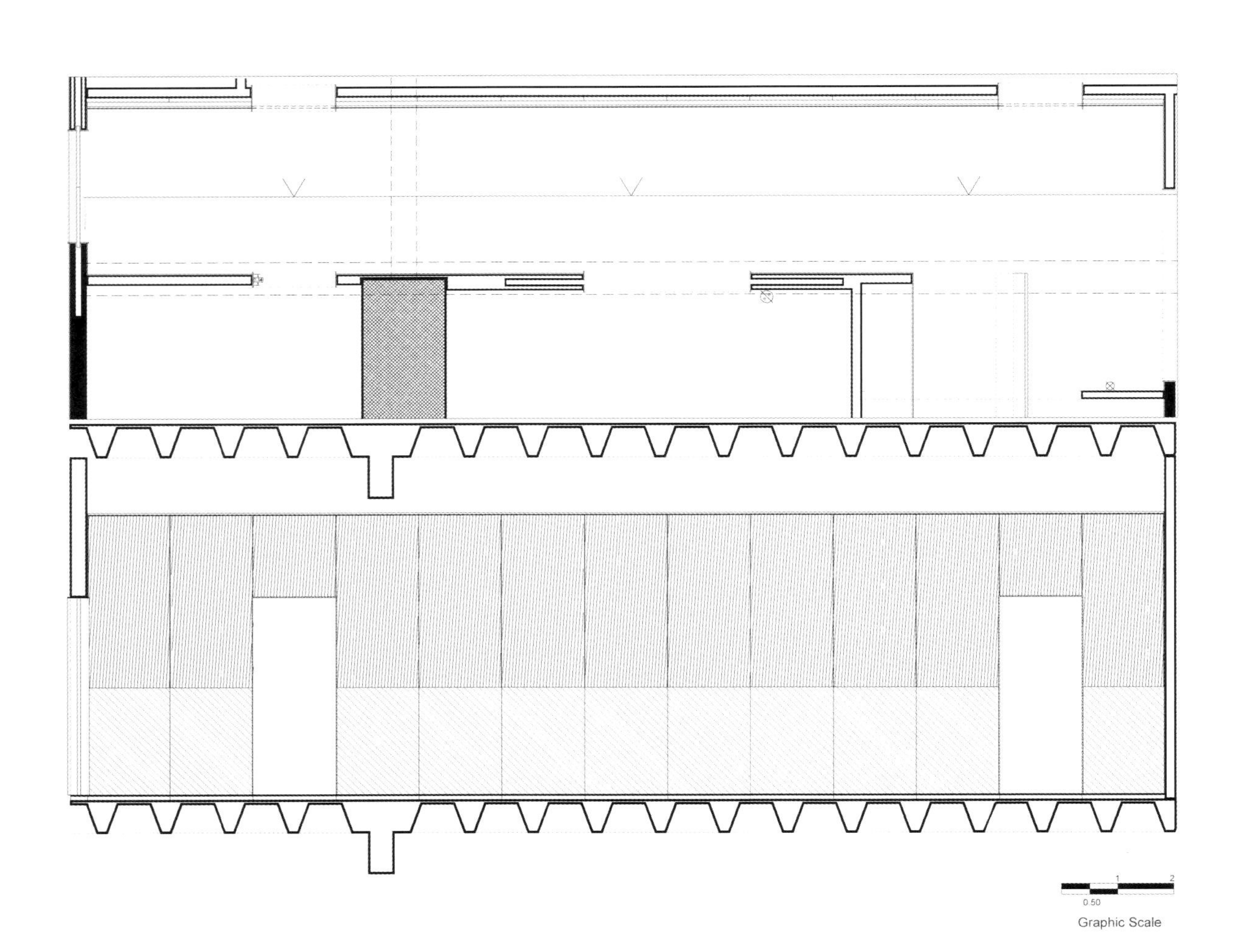
0.50
1
2
Graphic Scale

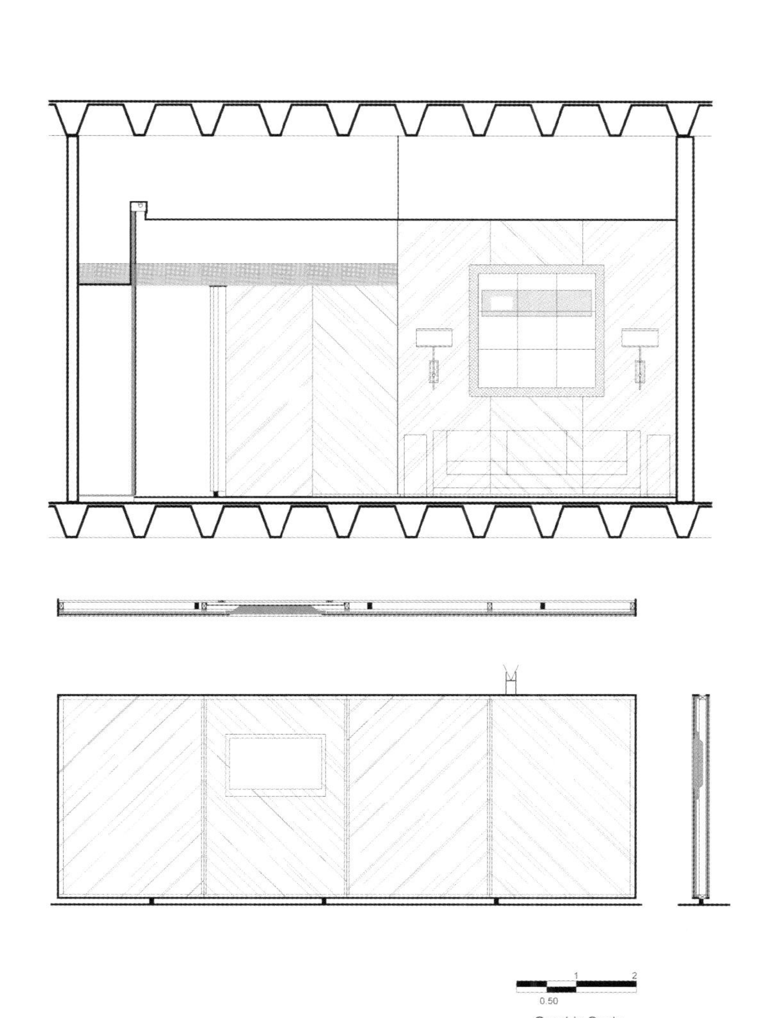
0.50
1
2
Graphic Scale

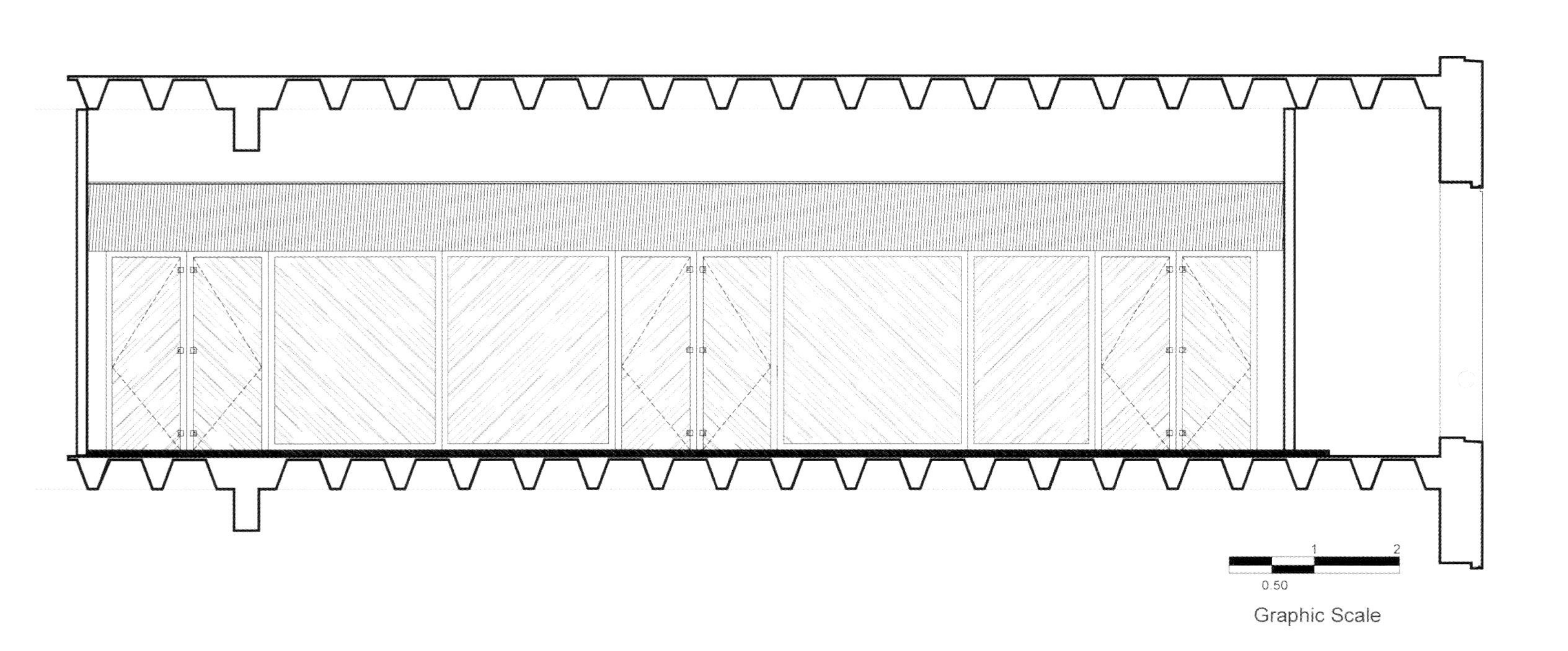
1
2
0.50
Graphic Scale

Office EXACTUSENSU

23

Design Company: CORREIA/RAGAZZI ARQUITECTOS

Project Location: Oporto

Area: 165 m^2

Photographer: Alberto Plácido

Uncommon: an informed engineer asks the architect, already friend because authored the interior decoration of his home, to remodel two commercial areas and its adaptation to the facilities of EXACTUSENSU - a leading company with the knowledge in the field of fire safety and in the emergency organization. A dull space, but well located, and a contained budget, formed the conditions framework, stimulants, towards its achievement. It was thought to operate only as designing furniture, which can be integrated in charismatically ascetic nature of the space created, or easily recycled elsewhere, in order to expand. The enclosure defined by white walls and floor, contrasts with areas of greater plasticity in laminitis and aluminum, a design for quality and accuracy that is required to refer to the criteria of project elaboration and studies carried out there today.

Building
Fire Safety
Systems

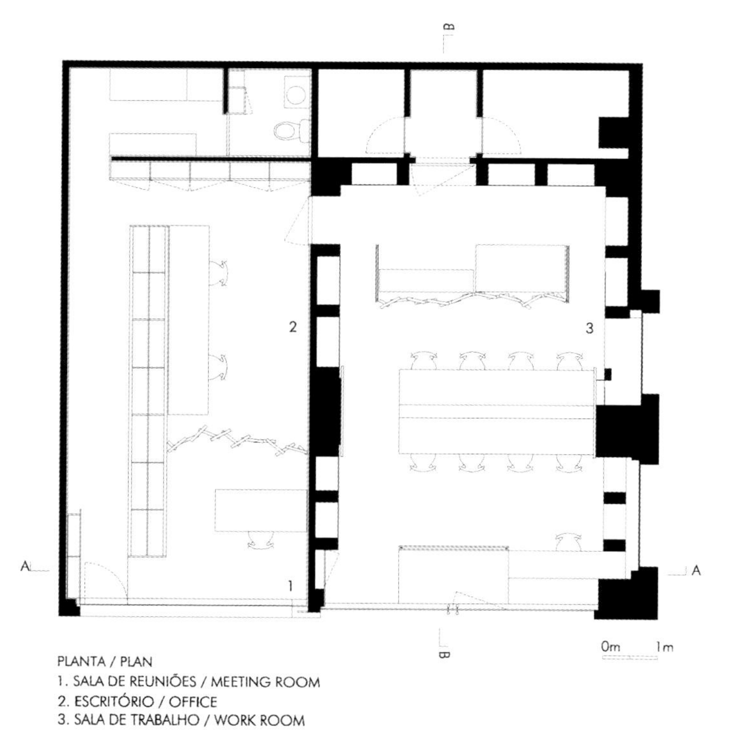

PLANTA / PLAN
1. SALA DE REUNIÕES / MEETING ROOM
2. ESCRITÓRIO / OFFICE
3. SALA DE TRABALHO / WORK ROOM

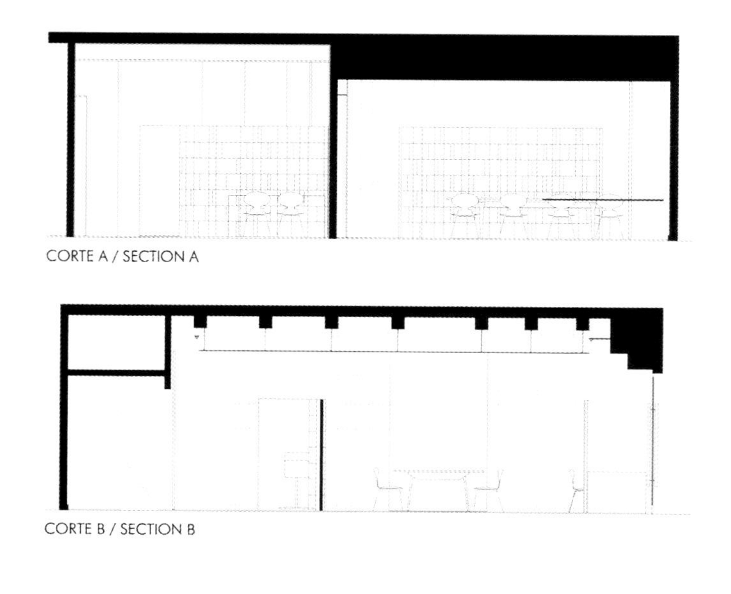

CORTE A / SECTION A

CORTE B / SECTION B

Kantar World Panel

24

Design Company: Space

Project Location: Condesa, Mexico City

Area: 1200 m^2

Photographer: Willem Schalkwijk

The designer was in a building on the Countess who needed to be rehabilitated with a moderate budget and reused most of the elements, so the design concept revolved around a dynamic, modern and sustainable above. The use of color was a fundamental part in the conception of this project which was retaken by the colors yellow and green of the brand to achieve the spatial change.

The reception is an industrial space with metal walls, screens and a yellow floor suggests that what happens inside the space.

When we entered office we have a different world, carpets, ceilings and walls with circulations in a mess at first glance is very tangible, but paying attention noticed the same order within chaos. A base color and paint and carpet accents totally changed the image of this space.

The design has an impact as a whole in three dimensions.

Large open spaces, a room type "crate" of reclaimed wood and a meeting room wrapped in a metal mesh facing the modern space and the flexibility to leave transformed over time from the hand of the company. On the 1st floor are the offices of management and a board room in white, black and yellow that involve the brand at every formal space.

The 2nd level is a large open space that serves as cafeteria, presentation and events as well as a large terrace where people can go off a bit of work.

All enclosed spaces, boardrooms, private phone and boots have their own name chosen by the people who work there, giving them a sense of belonging, commitment and brotherhood.

Color and lighting work together to improve the environment, productivity, sustainability and the final result.

Designed to exert little impact on the environment, these offices are green lines and low VOC's, lower energy consumption, green carpets, certified wood and recycling areas.

With all this was achieved to create a branding experience within the space of Kantar World Panel.

KANTAR WORLDPANEL
High definition inspiration
521065 64322 52
KANTAR WORLDPANEL
High definition inspiration

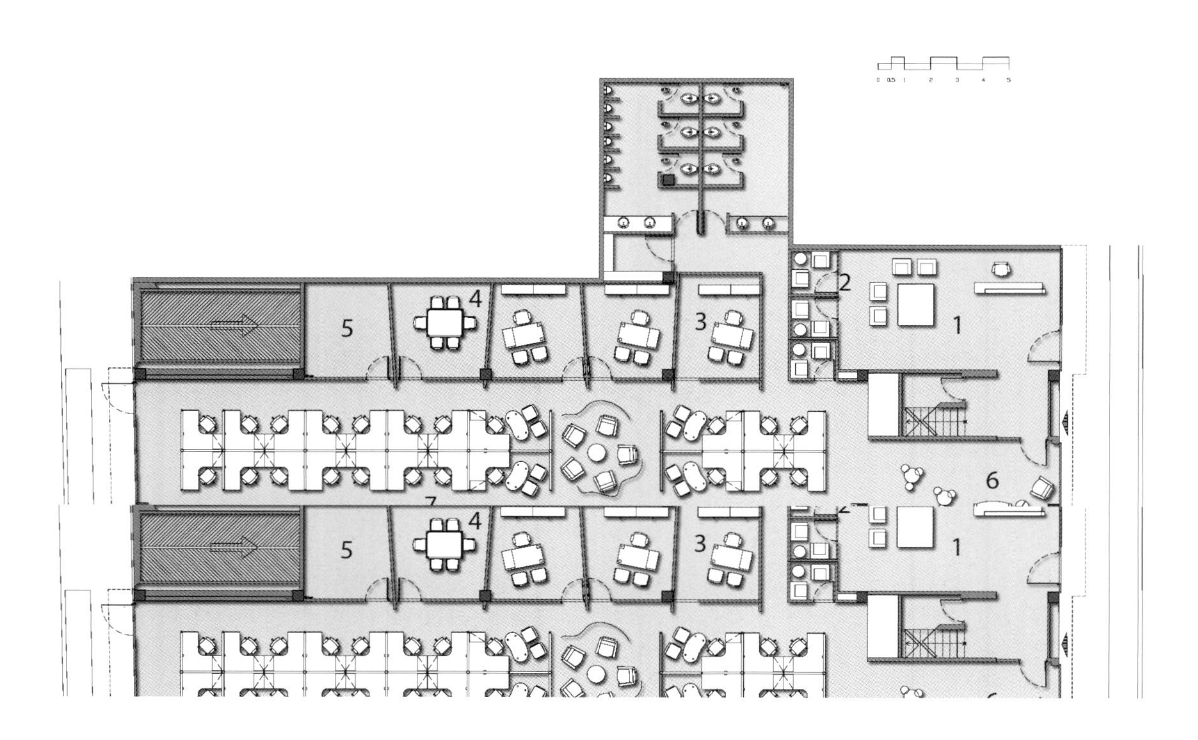
5
4
3
2
1
6
7

Jorge Dominguez
Venegas

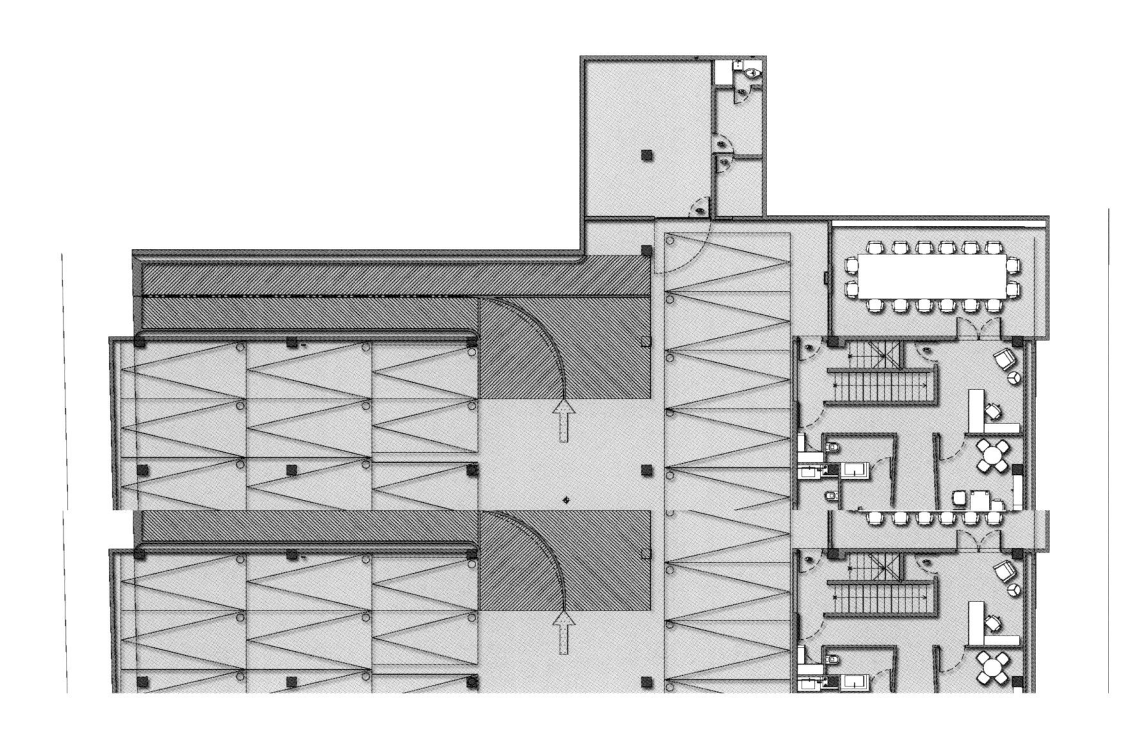

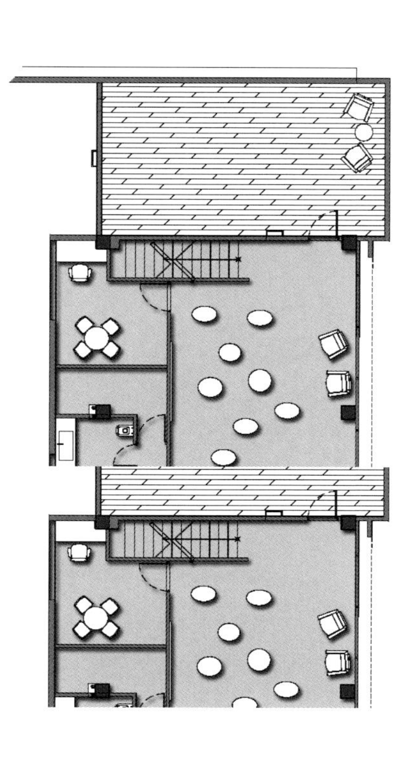

OFFICES IN TORRE MURANO

25

Design Company: Space

Project Location: Polanco,Mexico CITY

Area: 16,146 m^2

Photographer: Pim Schalwijk

This client needed to move to a more corporate space without losing that warm feeling they had achieved in their offices along time.

Space is divided in two areas, the technical part and the corporate side, joined by a zone formed by formal and informal meeting spaces. This is a transition that provides service to the whole office without doubling square footage.

The color use and combination was essential part in the design because we try to reflect the colors of nature with them to recreate the exterior feeling in the interior space. The use of blues, greens and oranges used together helps productivity and purify the ambience..

The general image is a hybrid between class and trend. A peaceful space with few private offices, a big open workspace and lots of alternative work stations, talk about the changes the company is making in how they work. New generations are taking the company towards the future so the space was designed to be dynamic and modern to comply with the demanding needs.

A sober, yet colorful, reception welcomes the office, allowing the user to glance the casual collission and phone booths behind the threshold.

Private offices and meeting rooms were located in the south facade in order to help reduce heat and help lower energetic costs.

This is an office designed to be as sustainable as possible, thought to be an ethical space for everyone.

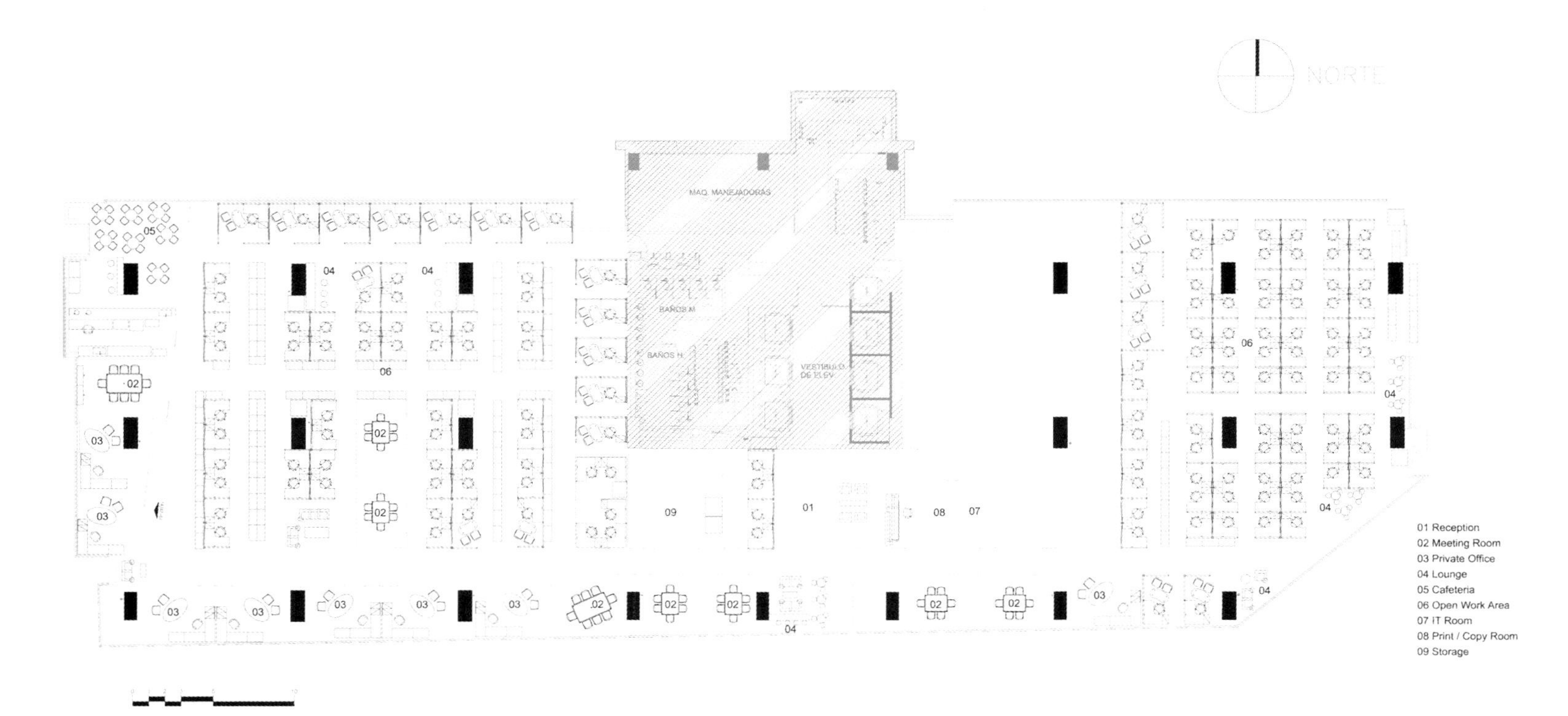
NORTE
MAQ. MANEJADORAS
BAÑOS M
BAÑOS H
01 Reception
02 Meeting Room
03 Private Office
04 Lounge
05 Cafeteria
06 Open Work Area
07 IT Room
08 Print / Copy Room
09 Storage

Prodigy MSN

26

Design Company: Juan Carlos Baumgartner LEED AP, Fabiola Troyo del Valle LEED AP

Project Location: Santa Fe, Mexico City

Area: 800 m^2

Photographer :Paul Czitrom

1.Open and bright as the future of Prodigy MSN.
The openness and translucency are evident in the whole Project. The layout gets all the advantage it can to the daylight. Most of the glassed façade is free, but there are a few points in which some meeting rooms and privet offices are next to the façade. Those spaces are closed with glass, that allowed the openness and brightness we were looking for. In addition, the workstations are low and they have translucent boards.
Finally, in the hart of the project was built a long piece of privets offices, conceptualized as a glass box in which it was designed some graphic patterns made of vinyl to made these spaces more privets but clear as well, they translate clearly the personality or Prodigy MSN.
2.Multi-propose spaces.
This project was focused to be versatile. So, if we compare this project with a traditional one, we can find that this one didn't look for a check list of needed spaces, moreover it creates spaces capable of respond to the client necessities. A clear example of this is the reception, which more than a wait-reception space, it is a node where employees, clients and visitors gets in. Into this node, it is located a coffee bar to make this approach warmer. There isn´t any physic board to separate this space to the work-space, and most of the meeting rooms are located into this node.
The offices have a variety of casual meeting rooms spread in strategic points to allow communication among employees, and some of them have other functions, as the collection and storage bar.
3.LEED Commercial Interiors certification.
The main points in this project for LEED certification are the following.
-The offices are into a building that is looking for LEED Core & Shell certification, and that building is located in an urban area that is provided by basic and accessible services and by public transportation infrastructure.
- The project saves more than 20% in the water use.
- The project gets a save of 20% of energy in lighting systems.
- The project complies with the standards of energy efficiency that LEED recommend.
-There is a dedicated space to collect and storage recyclable materials.
-More than 20% of materials used in the project have recycled content.
-The HVAC system has filters that help to maintain the interior air quality cleaner.
- Paintings, coatings sealers and glues used in the project have a low content of VOC´s, which avoids interior air pollution.
-100% of the workspaces have access to daylight and views.

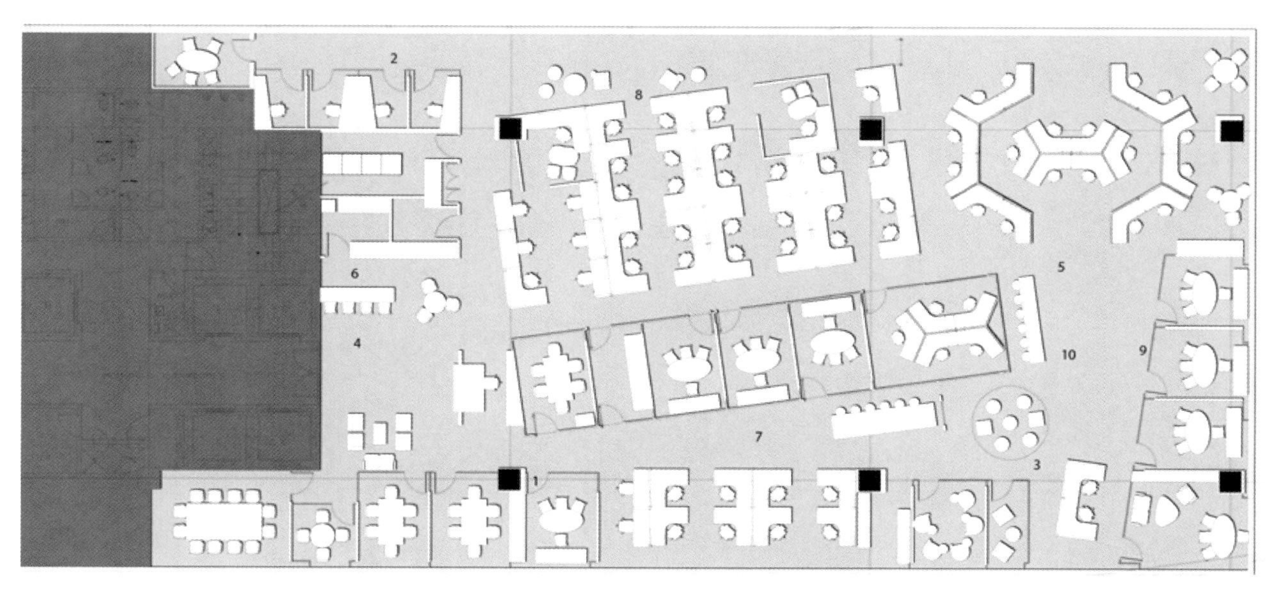

1. PRIVATE OFFICE
2. PHONE BOOTHS AREA
3. CASSUAL COLITION
4. RECEPTION
5. NEWS ROOM
6. COFFEE AREA IN RECEPTION
7. RECICLABLE WASTE STORAGE
8. OPEN OFFICE AREA
9. CASSUAL MEETING AREA
10. CASSUAL MEETING AREA

Government Officebuilding Utrecht

27

Design Company:Zecc Architecten BNA

Project Location: Netherlands

Area : 1100 m^2

Photographer:CornbreadWorks

Information

The internal alteration of an office Floor has created 70 to 80 places for working. The goal was to create a large variety of working environments that would match the activities and atmosphere on the office. The employees don't have their own working space but use the space that match their work at the time they need it. Spaces we designed are concentration rooms, temporary meeting rooms, standard working space, lounge working spots etc.

The exceptional areas are connected by a constructed element in a recognizable color. This architectural object gives the space an informal character and is made out of natural material. In the design we also integrated a green garden which we had designed by an artist.

Designing for the New way of Working in government

The vision of office space is constantly moving. Fixed workplaces are increasingly disappearing and give way to action-oriented, flexible workplaces. Even in government is blowing a fresh wind, which accommodate the new way of working. The tax formulates the change from fixed to flexible working as: organization-oriented housing. This new way of office accommodation saves the government approximately 30% office space, because part-timers, itinerant workers, leave and home working that in a typical office about 50% of workstations remains unoccupied! Besides economic reasons to pursue the tax authorities' commitment and inspiration "of (preferably young) employees. The average age of employees in the tax is now 49 years. Activity-based work promotes also communication on the workplace. It fits a less hierarchical organization and there is a greater knowledge exchanged. There is more space for (informal) meetings and people can feel at home and at work. Furthermore, this type of housing accommodate a specific way of working, which reflects the identity of the organization.

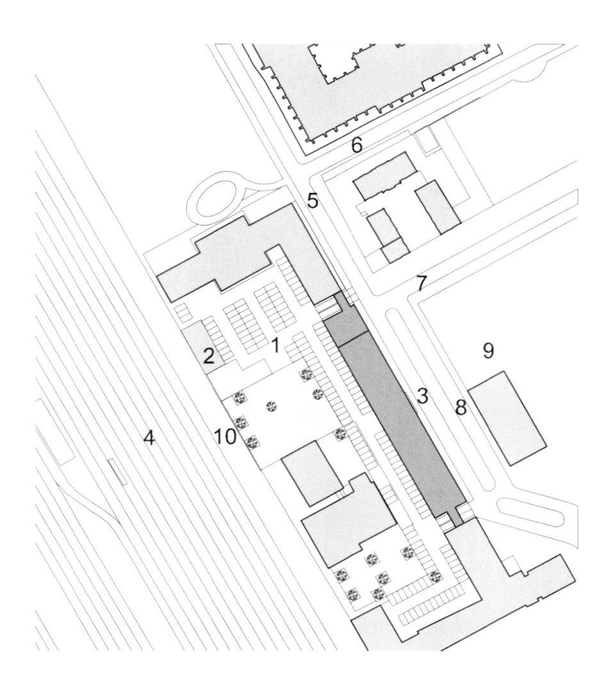

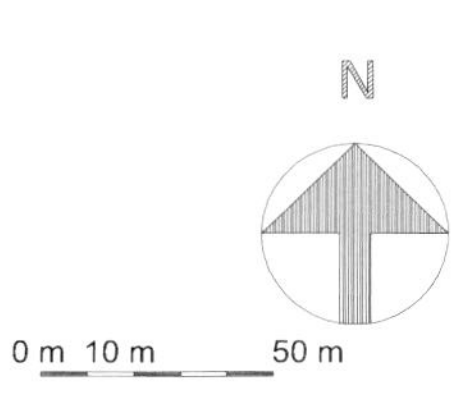

NL

1 parkeerplaats
2 fietsenstalling
3 hoofdentree
4 spoor
5 LAAN VAN PUNTENBURG
6 MOREELSEPARK
7 STERRENBOS
8 HERMAN GORTERSTRAAT
9 VROUWE JUSTITIAPLEIN
10 A V SCHELTEMABAAN

ENG

1 parking
2 bicycle parking
3 main entrance
4 railway
5 LAAN VAN PUNTENBURG
6 MOREELSEPARK
7 STERRENBOS
8 HERMAN GORTERSTRAAT
9 VROUWE JUSTITIAPLEIN
10 A V SCHELTEMABAAN

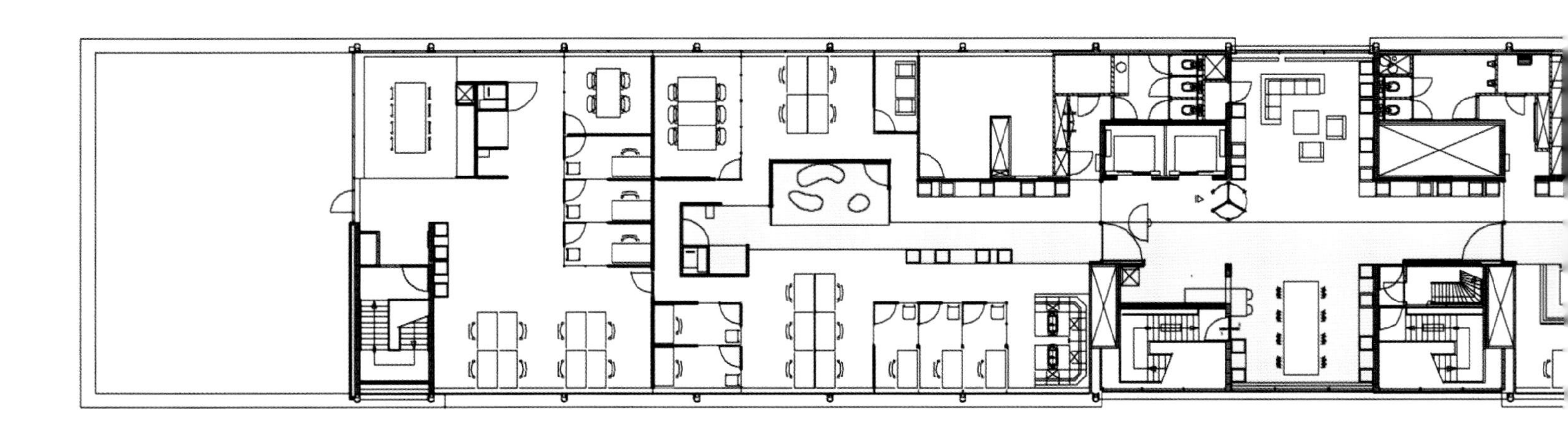

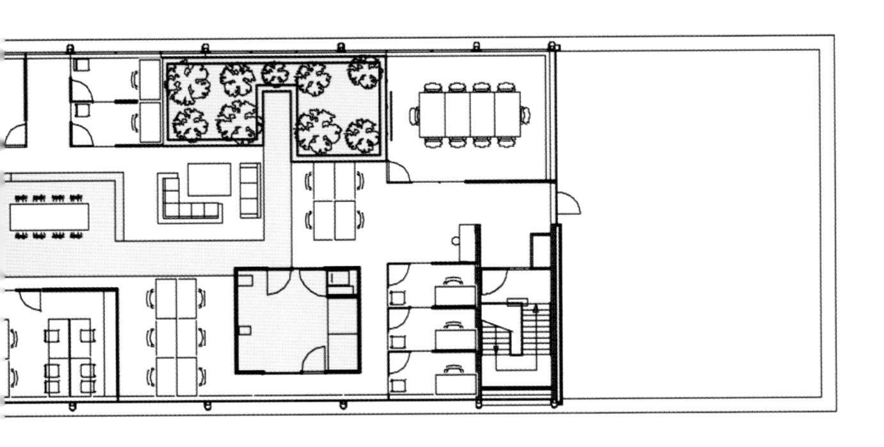

Glem Offices

28

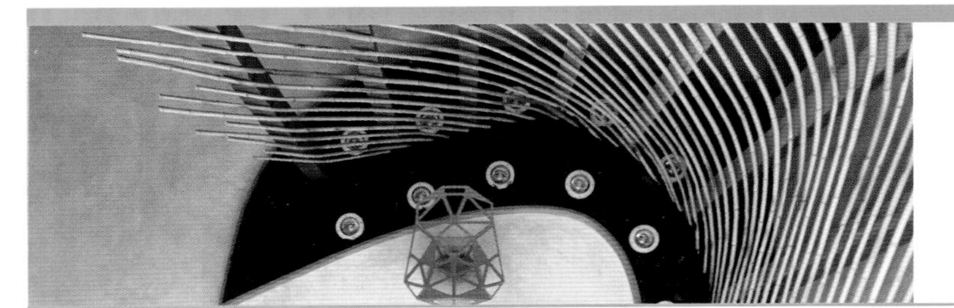

Design Company: Mareines + Patalano Arquitetura

Project Location:Lagoa, Rio de Janeiro, Brasil

Area: 35000 m²

Photographer: LEONARDO FINOTTI

The design of Glem offices in Rio de Janeiro, Brazil, is defined by it's odd location, under a triangular end of a concrete stand for rowing competitions. Besides the fact that the first of the three floors is semi-buried and the roof is public seating made of concrete, thus not allowing openings to the exterior, windows were extremely restricted on the only two small façades of the project. One is almost entirely occupied by a pedestrian ramp to access the stands, wich are public and so, seen by the client as a threat to a possible trespassing. For that reason, this façade has some long and narrow openings closed with clear transparent glass, to invite inside only light and external views. The second façade overlooks a group of confused existing buildings, a fact that led us to design a semi-circle glazing with translucent bullet-proof glass. After dealing with all that limitations, we decided to bring together all the separate parts of the project via a vertical circulation space defined by curved concrete stairs, eucaliptus laminated beams and a bamboo mesh. An empty 'basket' that welcome visitors and animate the office with a delicately enclosed void. On the first level, the basket-defined space offers a small waiting area used as well for informal meetings, and distribute access to three sinuous meeting rooms, two bathrooms, a working area and a technical room. The first staircase takes to a catwalk that mirrors, in laminated translucent glass, the semi-circle geometry of one of the façades glazing. This catwalk takes to the second floor that contains wood sliding doors that can partially connect or totally separate the spaces of the office's directors' rooms. The second staircase takes to the relaxing level of the office, with a small kitchen, a table for eating and a sofa for resting. There are also two changing rooms here. Details such as the on-site handcrafted bamboo mesh, the translucent catwalk glass, stainless steel handrails that seem to float over the stairs for not being attached to it, and the generous light filtered mainly trough the semi-circle glazing, gives lightness and delicacy to the design.

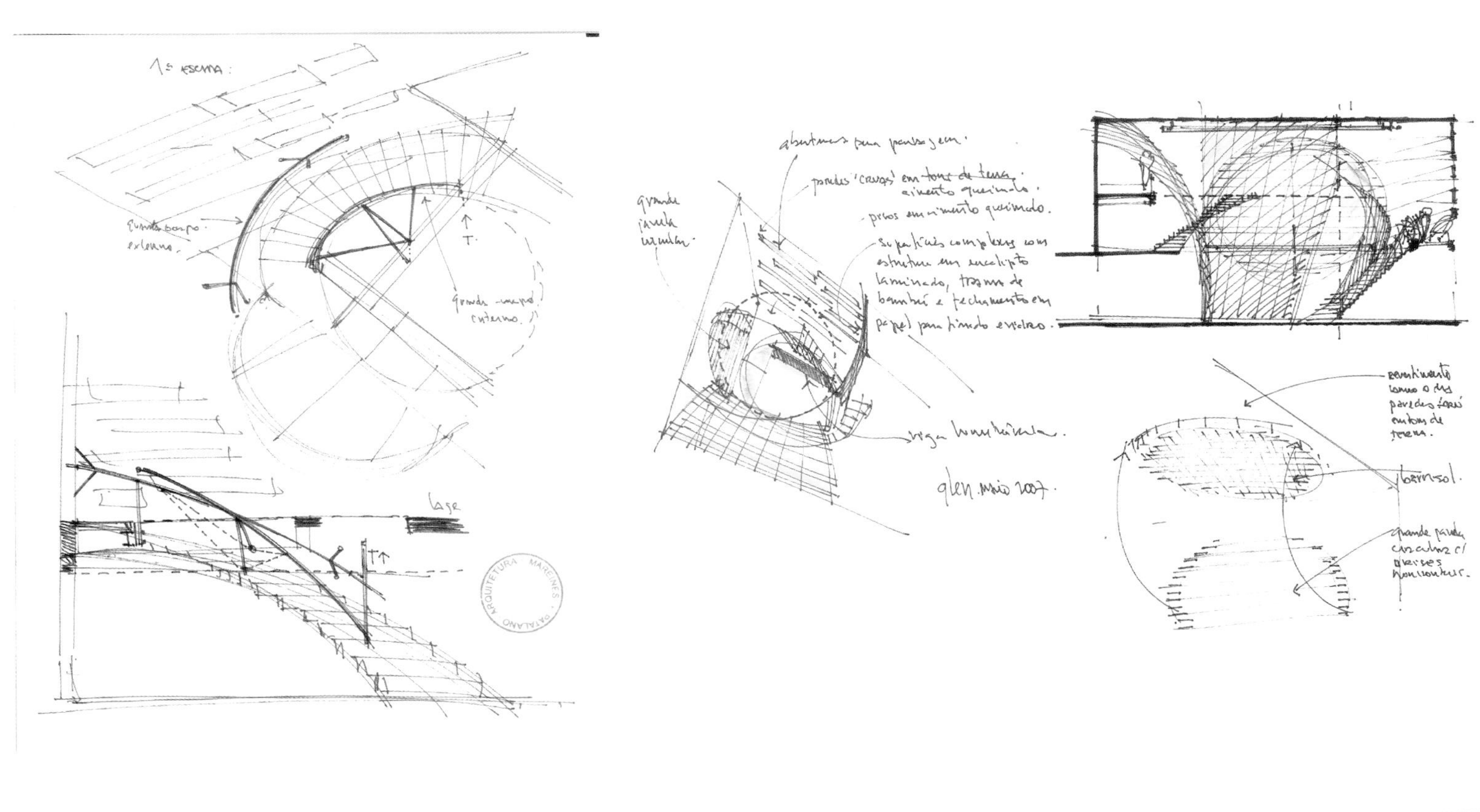

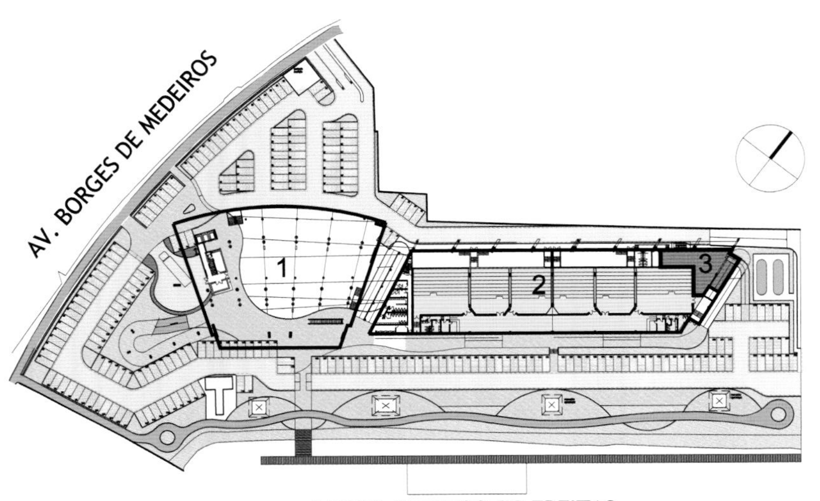

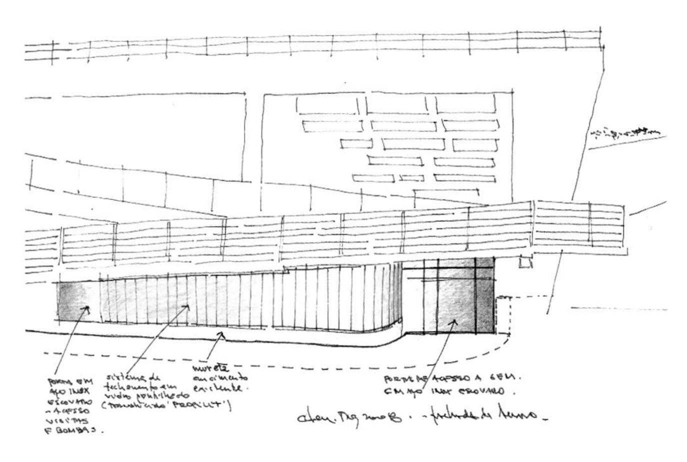

location / ground floor
1-Rowing stadium of Lagoa Rodrogo de Freitas
2-Grandstands to Pan american competitions of 2007
3-GLEM

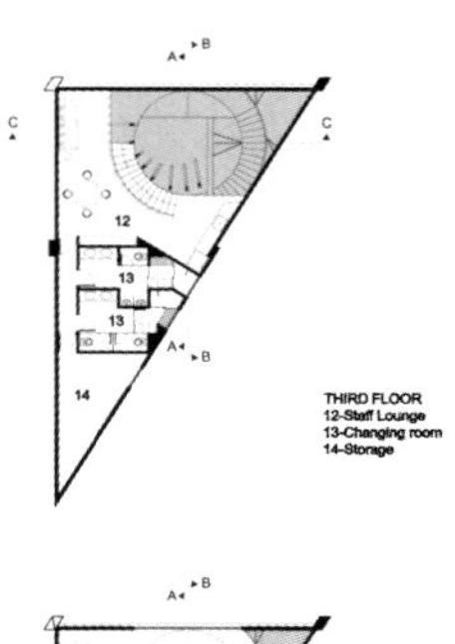

THIRD FLOOR
12-Staff Lounge
13-Changing room
14-Storage

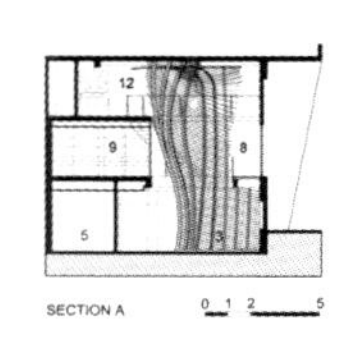

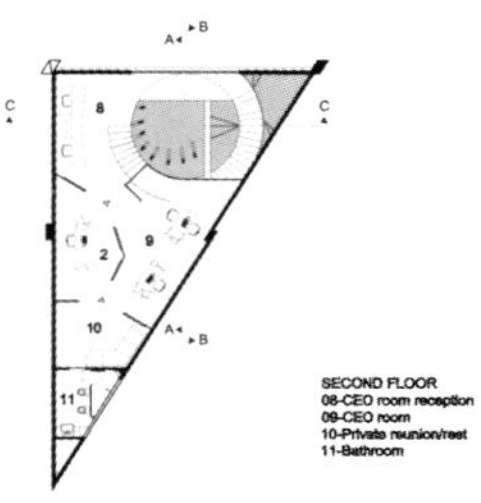

SECOND FLOOR
08-CEO room reception
09-CEO room
10-Private reunion/rest
11-Bathroom

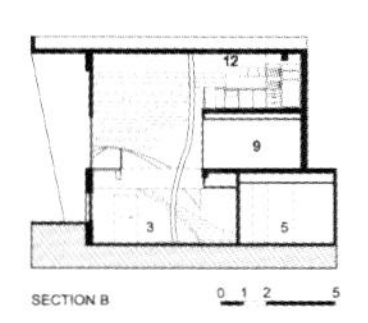

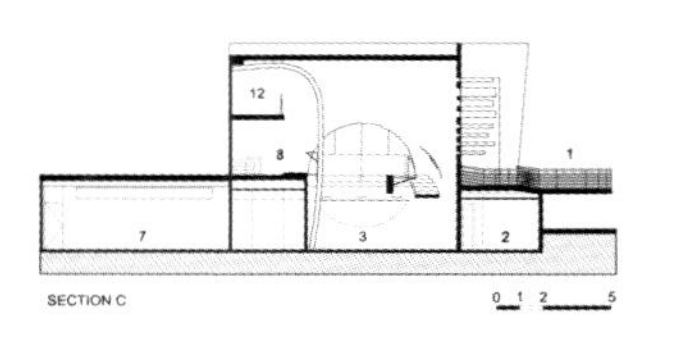

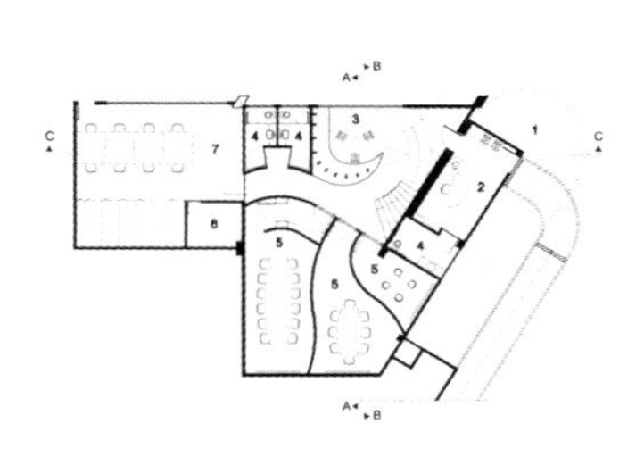

First Floor
1-Rowing stands acess (ramp)
2-Reception
3-Waiting area / informal meeting
4-W.C.
5-Meeting rooms
6-Audio and Vídeo room
7-Workstations

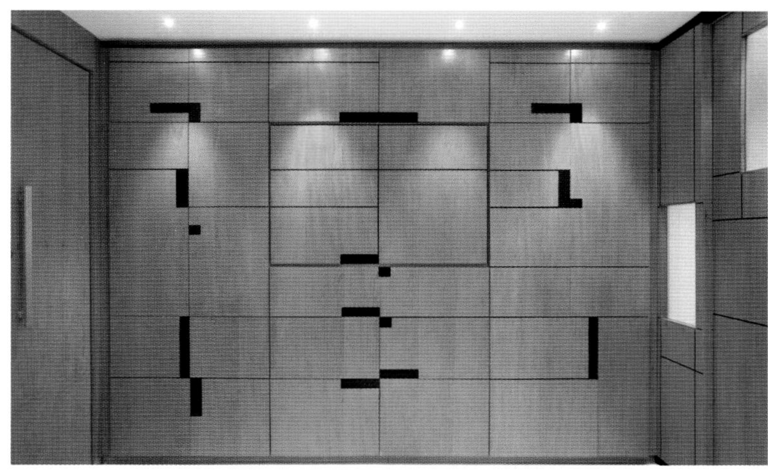

BAJAJ CORP OFFICE

27

Design Company:Collaborative Architecture

Project Location: Chenai, India

Area :390.19 m^2

Photographer:Lalita Tharani

The design for Bajaj Electrical, this Corporate Office is an attempt to take the project above the functional contingencies of a regular corporate work place brief, imbibing the space with an architectural character that would generate unparallel work efficiency.

The project had typical straight jacketed, hierarchical brief for a corporate office, with cabins for senior managers, cubicles for mid-level managers and workstations for executives, meeting room conference room, staff area and service areas. The brief was also emphasized to have orthogonal spaces in the office, in line with traditional Indian 'Vastu' principles.

Sustainability as an active strategy was fundamental theme right from the inception of concept design through selection of green rated products, active tools to cut down the energy consumption, day light harvesting and the general orientation of the work floor to tap maximum day light and reduce the south exposure to cut down cooling load.

The design challenge was to accommodate these highly debilitating guidelines, yet evolve a 'design narrative' which will cleverly mask the ubiquitous gridded plan.

Meeting room is placed strategically to segregate the public interface with the work interface. The contrasting dark color and the free flowing envelope of its space clearly generate a spatial boundary between Public-Work realms.

Enclosed spaces like senior manager cabins, conference, staff utility areas are positioned along one of the sides, allowing the workstations to be positioned to tap maximum day lighting.

The cabins have glazed partitions separating them from the workspace, allowing the day light to penetrate in to the cabin spaces. Collaborative designed the architectural graphics on the wall to ensure adequate privacy in the cabin area.

Every cabin, passages and entry foyers are provided with active tools to cut down the lighting load. The office is equipped with state of the art HVAC and BMS to make it truly energy efficient.

CONFER
CON
CONFER

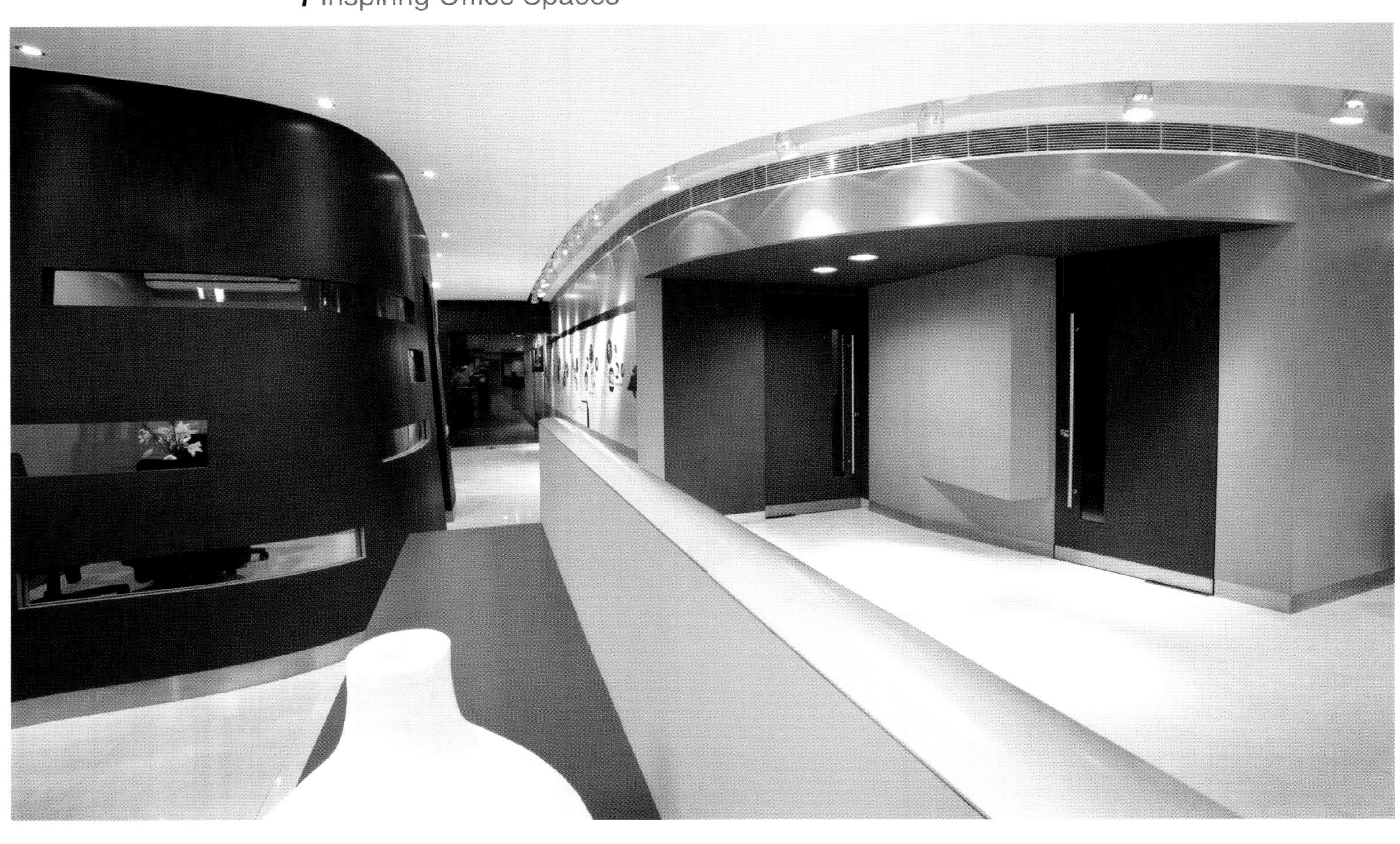

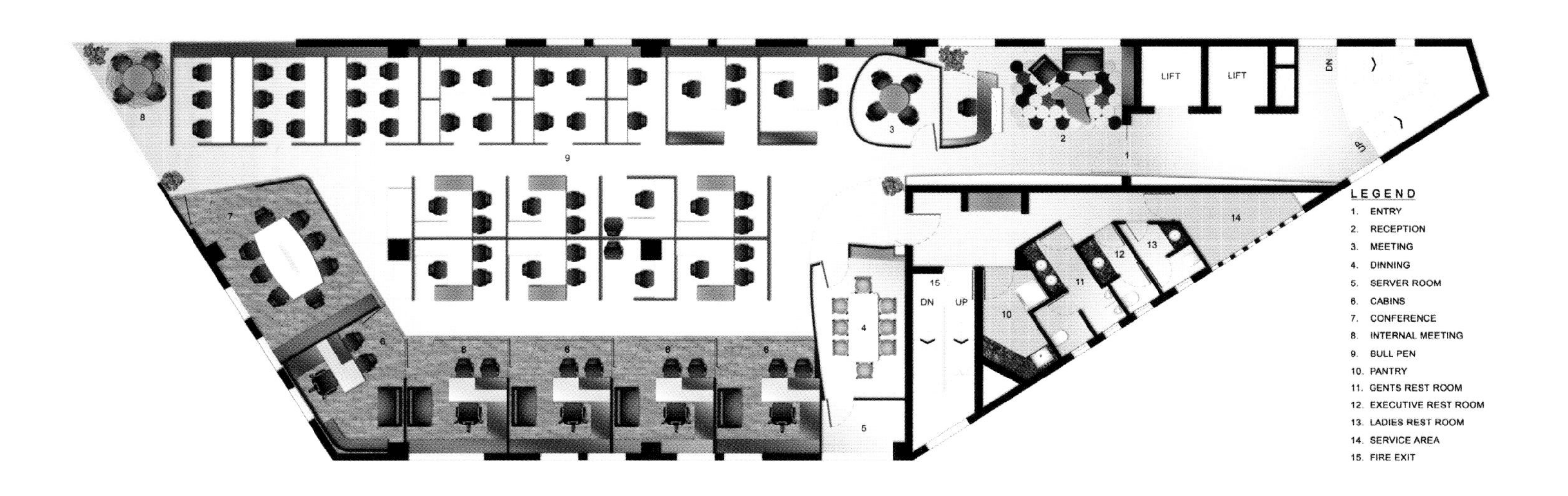
LIFT
LIFT
DN
UP
LEGEND
1. ENTRY
2. RECEPTION
3. MEETING
4. DINNING
5. SERVER ROOM
6. CABINS
7. CONFERENCE
8. INTERNAL MEETING
9. BULL PEN
10. PANTRY
11. GENTS REST ROOM
12. EXECUTIVE REST ROOM
13. LADIES REST ROOM
14. SERVICE AREA
15. FIRE EXIT

their quality of life while enhancing stakeholder

TWO

CONFER

GE Energy Financial Services Headquarters

30

Design Company:Perkins Eastman

Project Location: Stamford, Connecticut, USA

Area : 26 012.85 m²

Photographer:Paúl Rivera/ArchPhoto,Sarah Mechling Perkins Eastman

General Electric's (GE) Energy Financial Services Division is celebrated for its innovation in providing alternative energy sources for the United States and abroad. Their 280,000 sf headquarters relocation reflects this innovation by providing opportunities for consistent interaction between its team members.

This is a truly unique and healthy workspace for employees to work, interact, and share ideas. From visioning sessions with a cross-section of GE staff, and through the leadership of GE Energy Financial Service's CEO, a decision was made to shift to a 70% open plan work environment with significantly reduced private workplace. In addition, a huge shift was made in the provision of collaboration opportunities—which their CEO saw as critical to foster the ideas exchange the company depends upon. These collaborative venues included more traditional conference rooms and huddles, as well as a 50-foot-long white board with lounge seating under skylights as well as the "living room" a space with pantry, plasma screen and coffee bar to encourage chance encounters and ideas exchange. Recognizing the benefits that staff gets from health promoting programs and initiatives, GE also provided many amenities for its staff including: a state-of-the-art fitness center, storage and showers for bicyclists, classrooms for yoga, as well as jogging and walking trails around the headquarters property.

As a purveyor of alternative energy strategies, it was important for the project to incorporate leading environmental initiatives. Perkins Eastman worked closely with GE at the project's inception to find ways to make the new headquarters as environmentally responsible as possible. The building achieved LEED Gold certification for its use of sustainable products, lighting efficiencies, and other environmental practices. Perkins Eastman teamed with Sustainable Design Collaborative (SDC) as LEED consultant to develop a strategy that included 100% use of green power from wind, 30% reduction in water usage, 20% use of recycled materials in the project's interior finishes, 20% use of materials manufactured within a 500-mile radius, "cradle-to-cradle" furniture, use of low volatile organic compounds (VOCs) emitting materials, and incorporation of alternative transportation, including provision for bicyclists. More than 75% of waste was diverted from landfills during construction. Innovative green design features also include a series of branding walls that explore the company's history and educate visitors and staff.

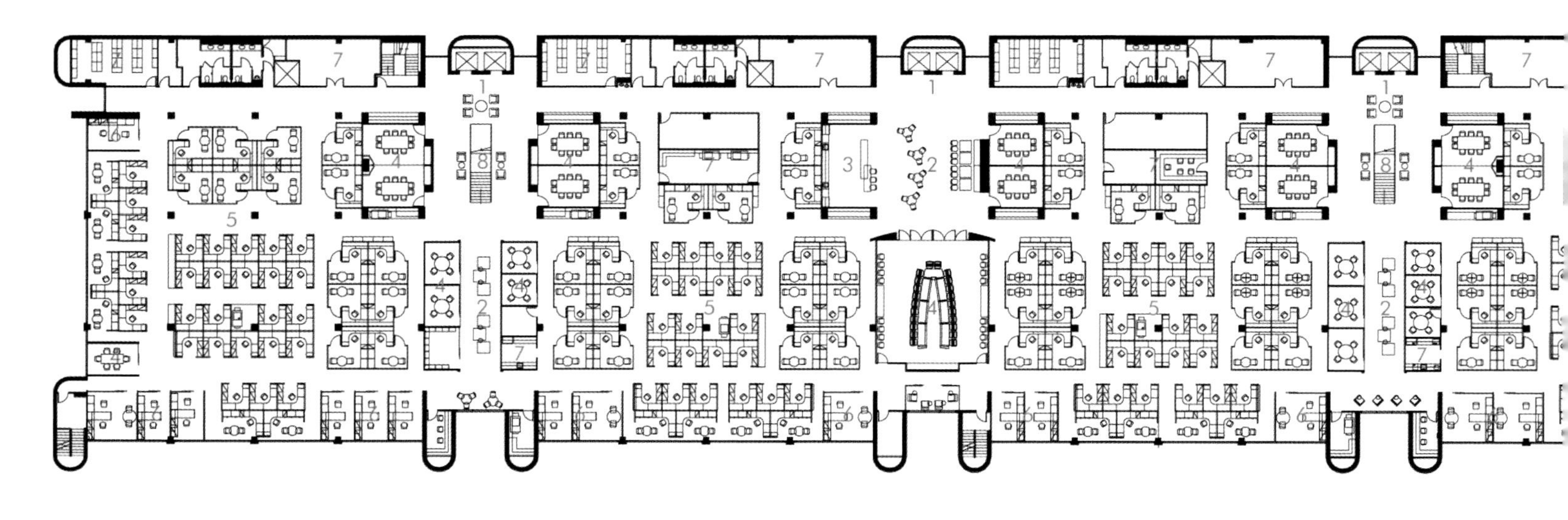

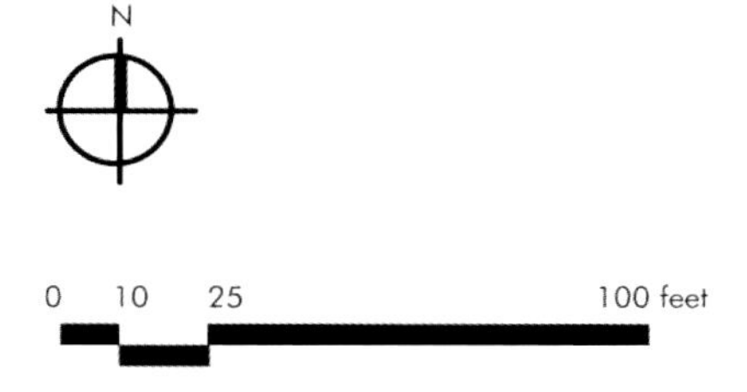

GE ENERGY FINANCIAL SERVICES typical floor

1 Elevator Lobby
2 Open Collaboration
3 Dining
4 Conference
5 Open Pla
6 Private O
7 Support
8 Commun

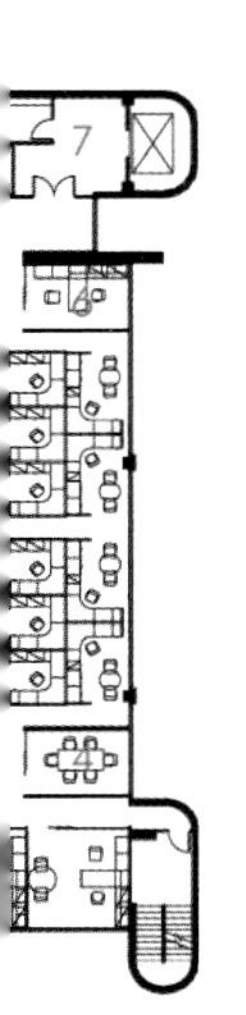

EDISON
POWER
GREEN
HEALTH

VictorinoxSwiss Army Brands, Inc

31

Design Company:Perkins Eastman

Project Location: Monroe, Connecticut, USA

Area: 14307.06 m^2

Photographer:Woodruff/Brown

When Victorinox/Swiss Army Brands, Inc. decided to expand their presence in the US they turned to top international architecture and design firm Perkins Eastman to find a suitable location for the company's future headquarters and to design it as well.

The client wanted to pay homage to the company's heritage by incorporating a clear sense of balance and proportion—the same virtues of Swiss design that have guided the company's own success for more than a century. The main entrance of the 120,000 sf facility features a two-story lobby with a product showroom, reception area and an interactive history wall. The company's own signature shade of red and custom light fixtures bearing their corporate logo are incorporated throughout the facility and subtly reinforce the company brand in concert with several other finishes selected from the company's own retail showrooms—paneled walnut and oak, glass, steel and stone. Several of the finishes—chosen for their functional versatility—prescribe to the "Cradle to Cradle" process and reinforce the utilitarian nature of the company's own product line in a contemporary and timeless aesthetic.

The new space needed to foster interaction among two unique groups of employees: administration and distribution that required a functional separation in the facility's design. To facilitate greater connectivity, shared common areas—cafeteria, lounge, staff gym and lockers—are centrally located between the warehouse and office spaces providing informal areas for the staff to socialize and interact. The facility also needed to support the company's own horizontal corporate hierarchy where every employee, regardless of title, is treated equally. Workstations are placed around the perimeter of the office space while offices for senior management are more centralized in the configuration within the space. Doing so broke up the "corporate alley," of conventional office design reinforcing the concepts of equality and respect.

Perkins Eastman created a new environment worthy of the company's venerable brand that honors its history and its heritage. Such a cultural crossover led to a design the combines the richness of modern and traditional elements in a timeless and contemporary setting.

VICTORINOX SWISS ARMY

VICTORINOX/SWISS ARMY BRANDS
floor one
1 Entrance
2 Lobby
3 Conference
4 Dining
5 Open Office
6 Private Office
7 Support
0 10 25 50 feet
N

EXIT

2061

Toto ish 2011

Design Company:MACH Architektur GmbH

Project Location: Frankfurt, Germany

Area : 1300 m^2

Photographer:Fotodesign Schiemann

For the second time MACH Architektur is assigned by the Japanese firm TOTO to design a fair booth of 1300 m2 (800 m2 in 2009) for them to represent the firm in Europe at the sanitary fair ISH 2011 in Frankfurt.

In order to reflect TOTO`s environmental-friendly philosophy, MACH Architektur designs a natural environment, which forms the suitable framework to exhibit the products in. The booth consists of a garden area surrounded by three open pavilions. Natural materials combined with modern architecture and the high-tech as well as high-end products from TOTO form a harmonious booth, inviting visitors to stroll around in the garden and discover the different product series inside the pavilions.

The rustic stone flooring together with the wooden terrace evoke images of traditional Japanese baths known as Sento and Onsen. With these as the visual background the ceramic products by TOTO and the clear lines of the furniture stand out and also show that TOTO is a firm with tradition, established in 1917.

TOTO

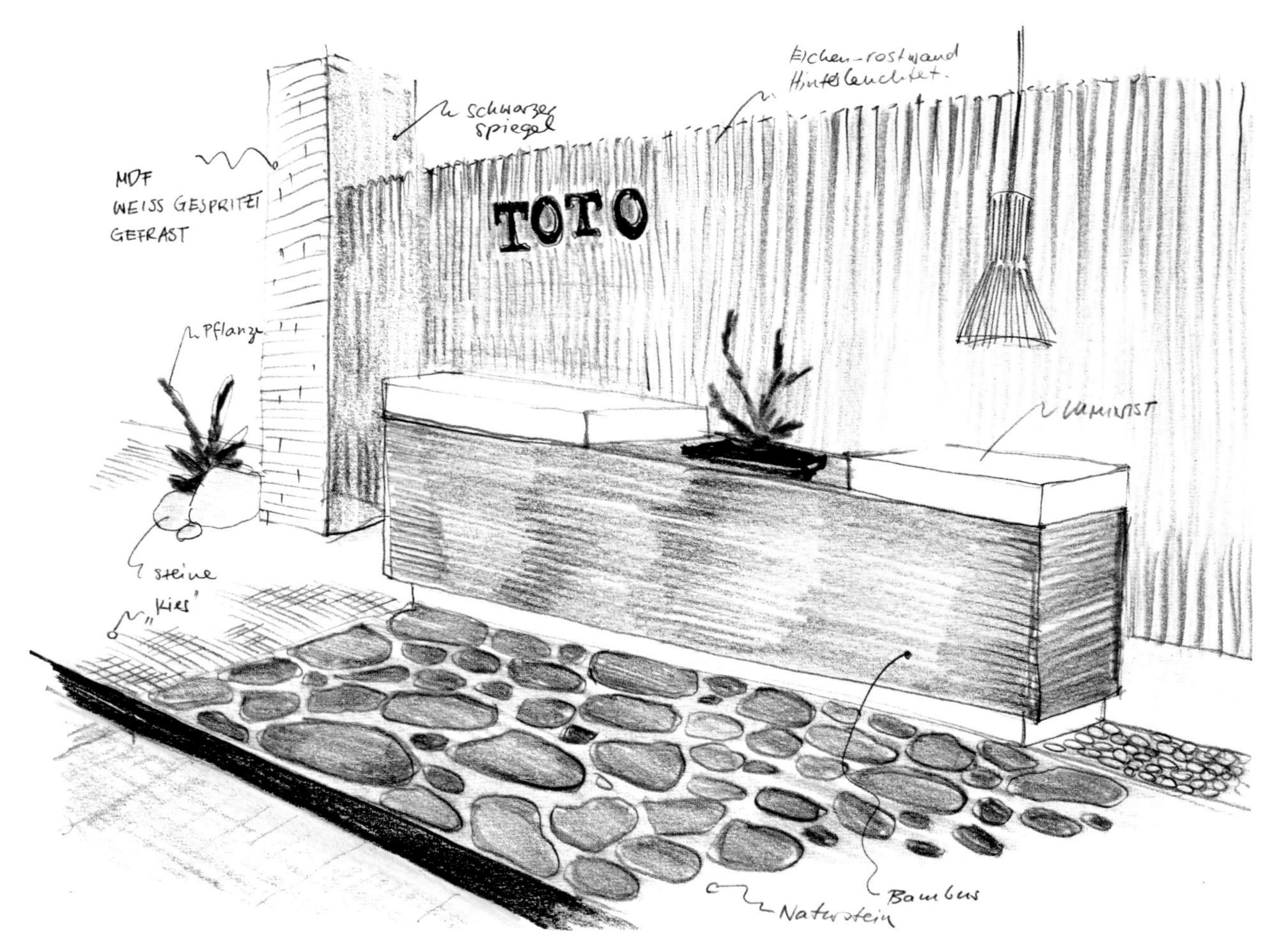
schwarzer spiegel
Eichen-rostwand
MDF
WEISS GESPRITET
GEFRAST
TOTO
Pflanze
steine
„Kies"
Naturstein
Bambus

TOTO
NEORESTʼLE

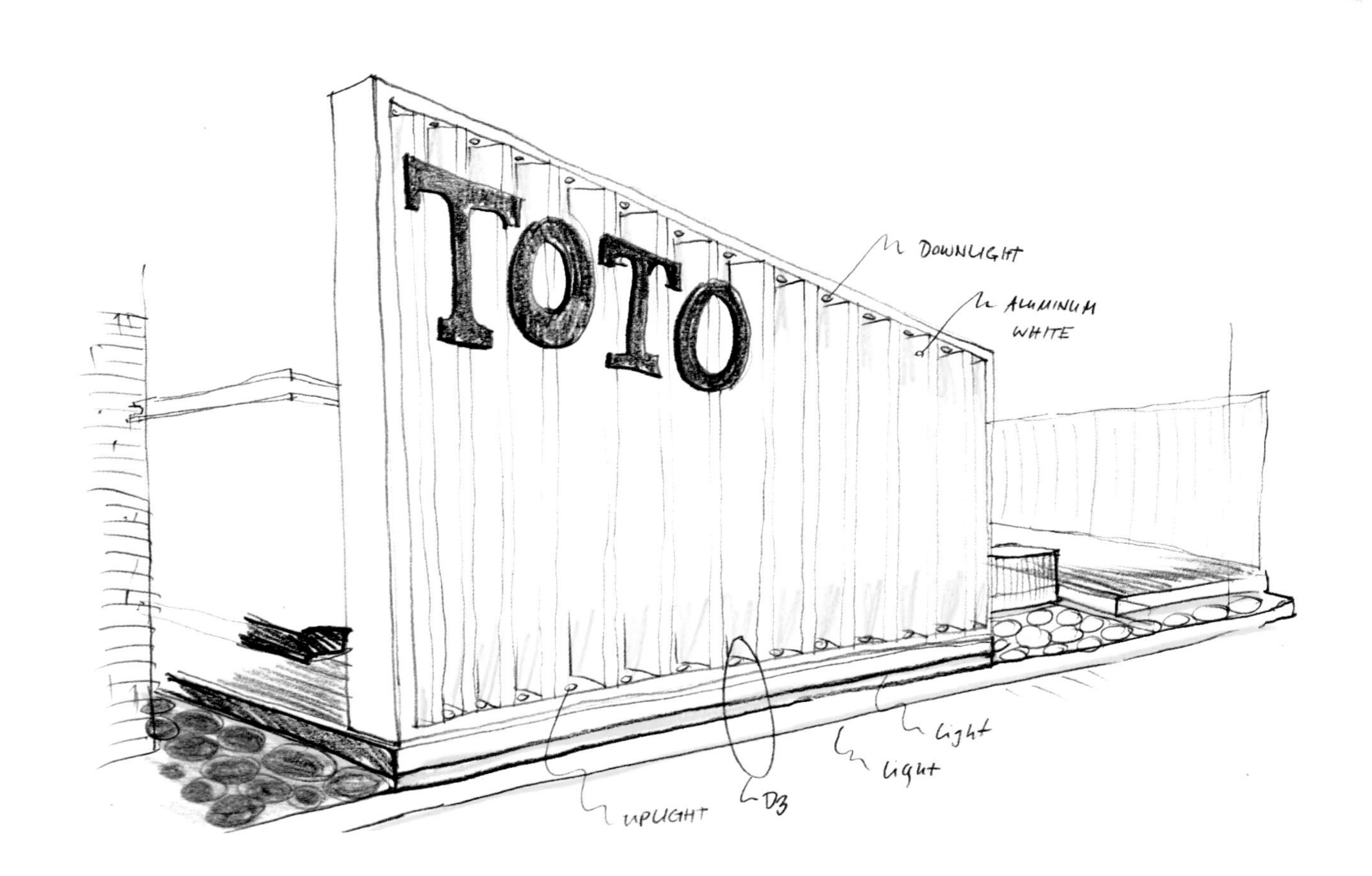
TOTO
DOWNLIGHT
ALUMINUM WHITE
light
light
UPLIGHT
D3

TOTO
VIP
LOUNGE
000 000 WASHLETS
ORLDWIDE

TOTO
TOTO

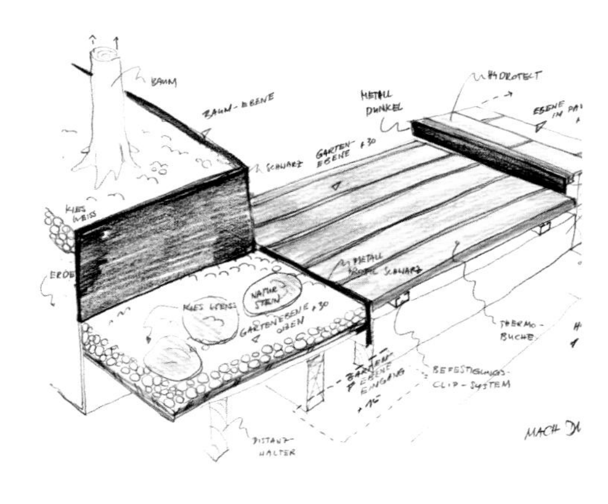
BAUM
METALL DUNKEL
GARTENEBENE
MACH DU

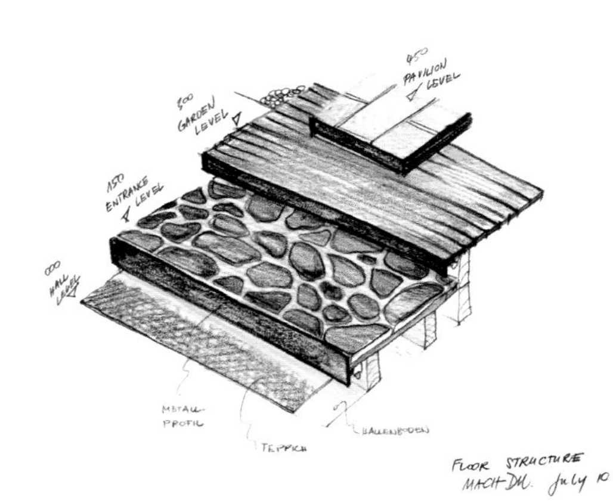
450 PAVILION LEVEL
300 GARDEN LEVEL
150 ENTRANCE LEVEL
METALL PROFIL
TEPPICH
FLOOR STRUCTURE
MACH DU. July 10

MACH DU
July 10

MACH DU July 10

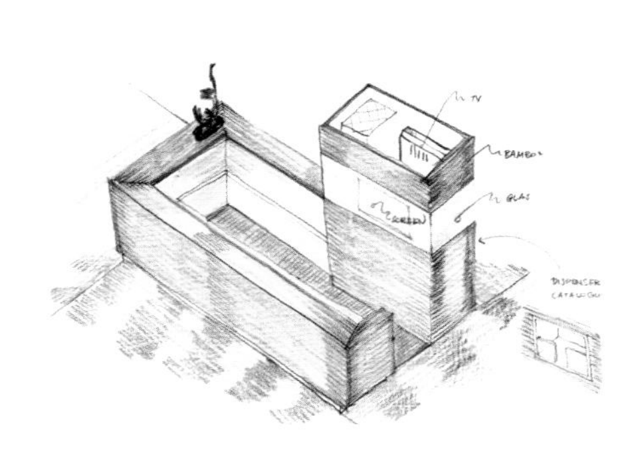

PUBLIC
SERIES

NEOREST SE

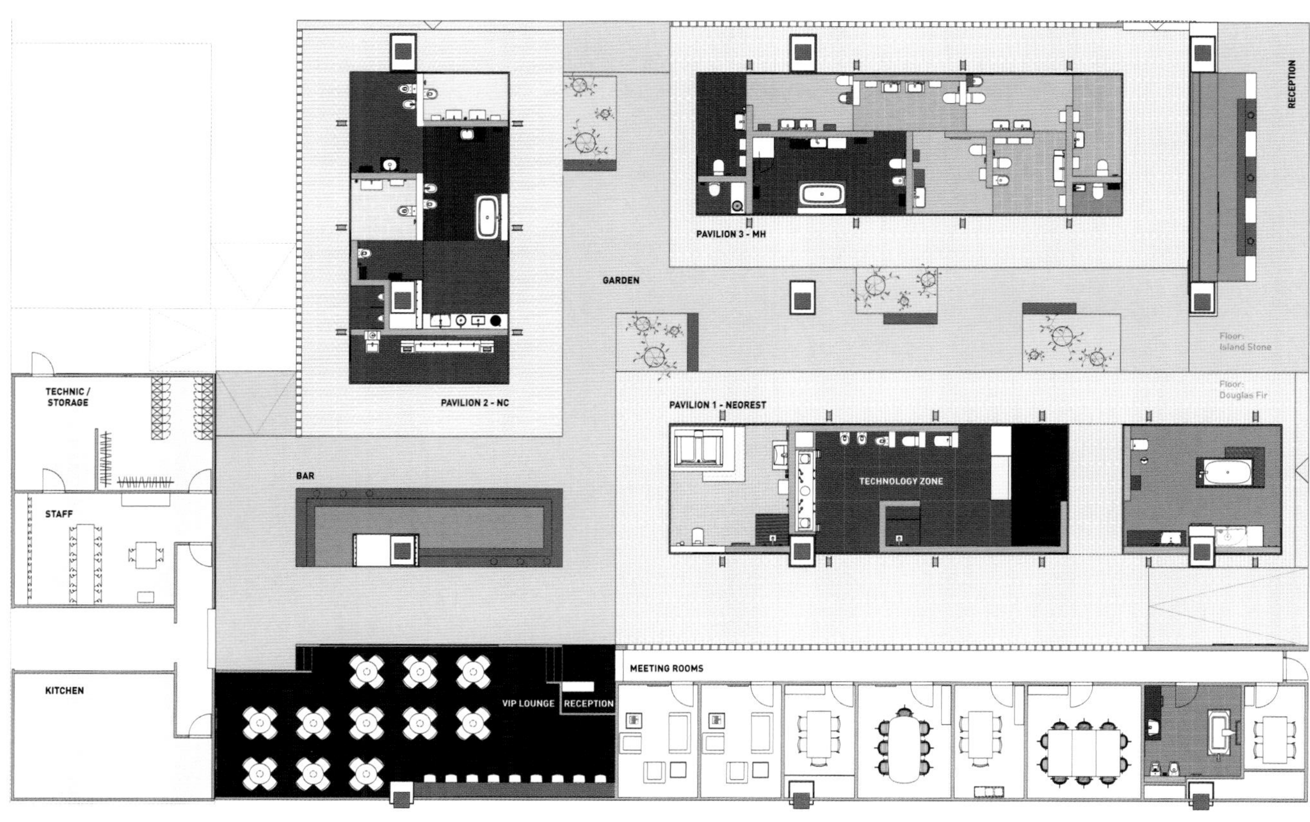
RECEPTION
PAVILION 3 - MH
GARDEN
Floor: Island Stone
Floor: Douglas Fir
PAVILION 2 - NC
PAVILION 1 - NEOREST
TECHNOLOGY ZONE
TECHNIC / STORAGE
BAR
STAFF
KITCHEN
MEETING ROOMS
VIP LOUNGE
RECEPTION

Cogeco Headquarters

33

Design Company:waltritsch a+u, Arch. Dimitri Waltritsch

Project Location: TRIESTE, ITALY

Area : 600 m^2

Photographer:Marco Covi

The City of Trieste is among the world most important port within the coffee trade market. Cogeco is a firm dealing as intermediate between the raw good and the coffee roasting plants. The project consist in the interior renovation of the company siege, underlining the two distinguishing factors which characterize the firm: the worldwide commercial relationships and the fact that Cogeco provides specific knowledge and lab test. These points have been particularly enhanced in the entrance lobby and in the proof and taste laboratory room.

The lobby of the Cogeco siege is characterized by a multi-layered folded wall, an abstracted map where only the parallels of the globe have remained, which hosts a series of "exotic" coffee names coming from all over the world, and by a big chemistry formula of the caffeine, which marks the corner and gives evidence of the specific knowledge. In such a way, one is immediately transported in a quick ride through the globe, and at the same time given a clear statement of the company's know-how.

The proof and taste lab is the place where the company makes a series of test's on the raw good in order to provide a certificate of quality (roasting, checking dimension and smell, tasting etc), and where most of the commercial deals are made. Coffee sample bags are exposed on a colored shelf marking the perimeter of the room, while one of the proof tables hides the 'spitting pots'. The space is again dominated by a big map folding on this table, a very detailed Goode Homolosine projection of the world, where precise geographic indication about the origin of the most important caffee types are given, adding a visual layer to the proofing experience.

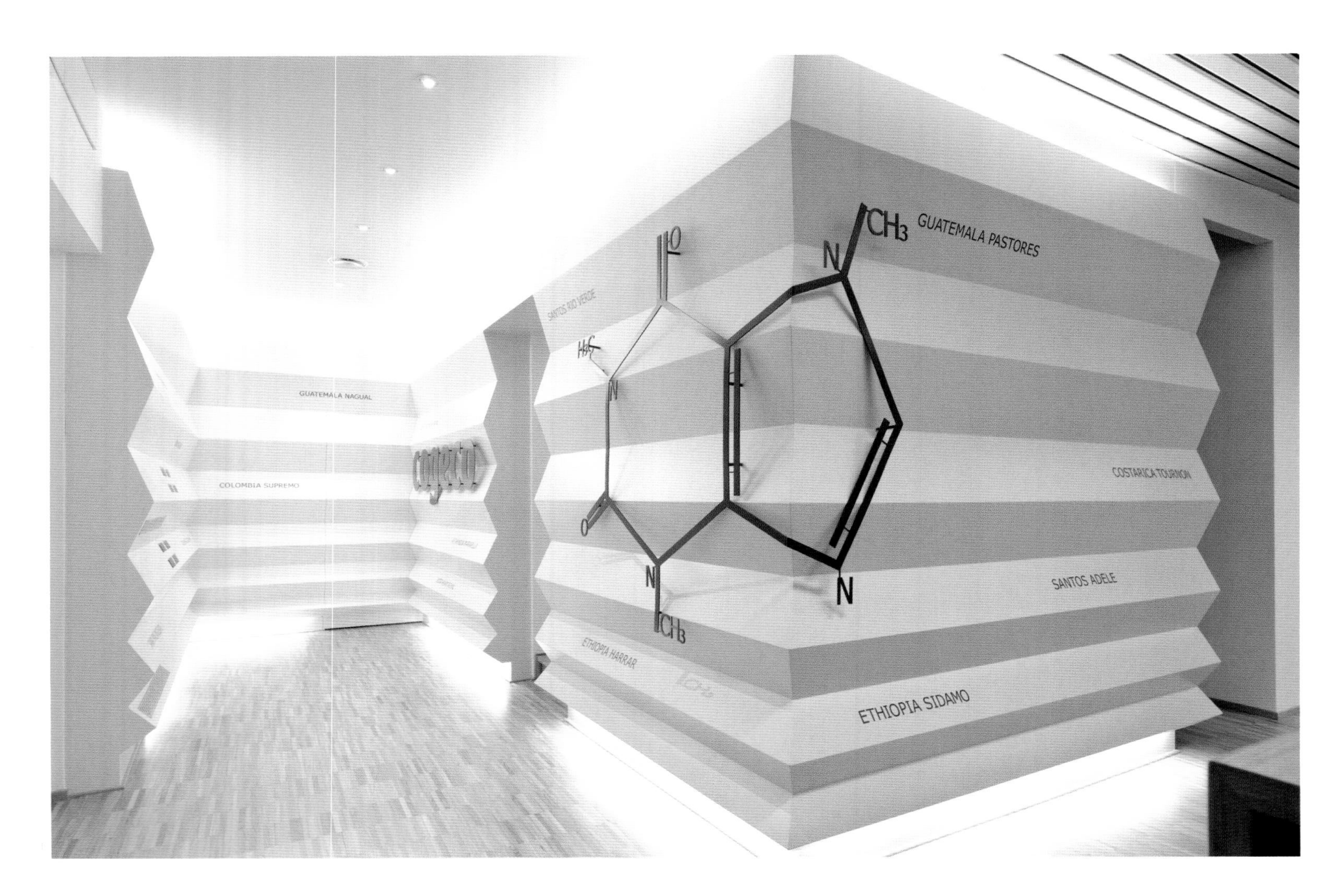

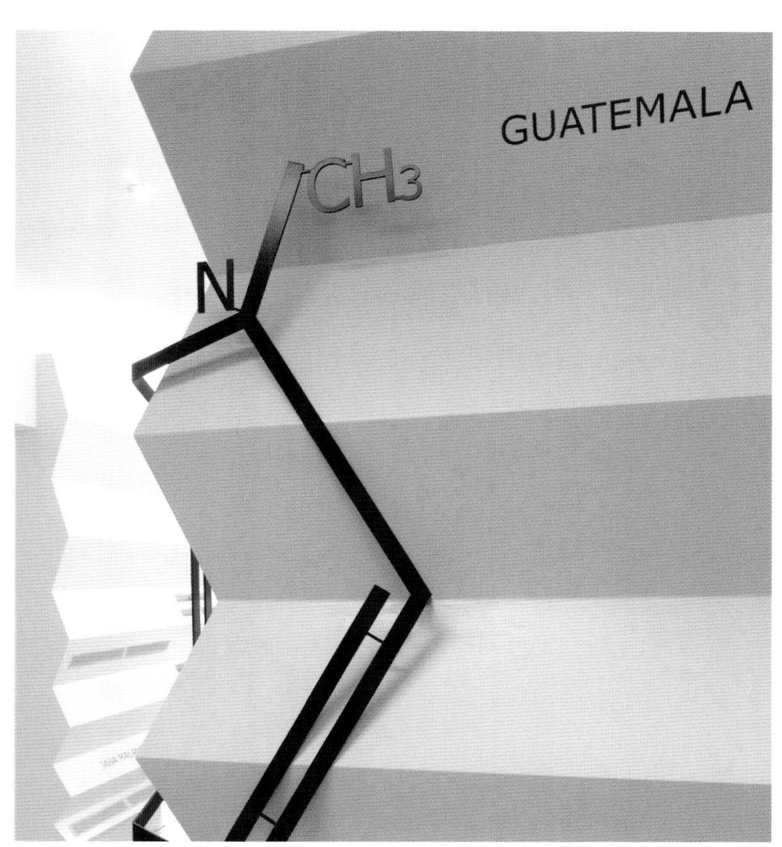
GUATEMALA
CH3
N

GUATEMALA NAGUAL
PORTO RICO SAN PEDRO

PORTO RICO SAN PEDRO
Cogeco
NICARAGUA MARAVILLA
INDIA KAAPI ROYAL
SANTOS RIO VERDE

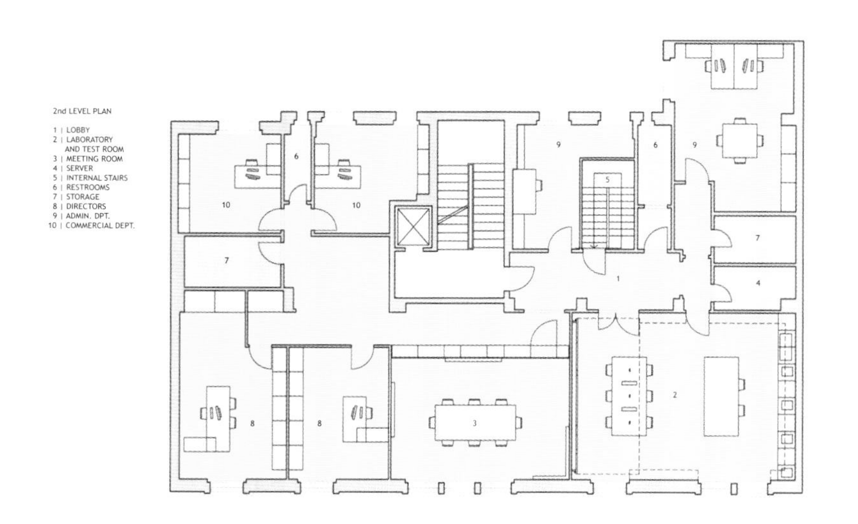
2nd LEVEL PLAN
1 | LOBBY
2 | LABORATORY AND TEST ROOM
3 | MEETING ROOM
4 | SERVER
5 | INTERNAL STAIRS
6 | RESTROOMS
7 | STORAGE
8 | DIRECTORS
9 | ADMIN. DPT.
10 | COMMERCIAL DEPT.

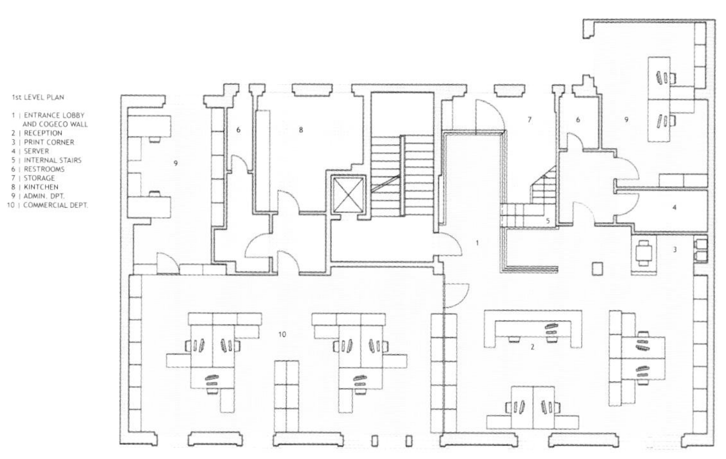
1st LEVEL PLAN
1 | ENTRANCE LOBBY AND COGECO WALL
2 | RECEPTION
3 | PRINT CORNER
4 | SERVER
5 | INTERNAL STAIRS
6 | RESTROOMS
7 | STORAGE
8 | KINTCHEN
9 | ADMIN. DPT.
10 | COMMERCIAL DEPT.

Uniflair

34

Design Company:Mario Cucinella Architects

Project Location: Conselve

Area : 3600 m^2

Photographer:Jean de Calan,Marco Covi

The Uniflair project consists of two interventions: the first concerns landscaping, seen as a patchwork that contaminates the mainly industrial character of the area.

As the site is in the countryside, the intention was to avoid the idea of industry "stealing ground" from the landscape, adding pleasant elements far removed from the uniform greyness of cement and asphalt: areas in red concrete, trees scattered so as toseem pre-existing, and green areas, such as a bamboo garden and a flower-strewn field.

The second intervention concerns the interior layout of the building allocated to the technical offices. The interior is a 200 m longopen plan where the various workplaces are brought together around a single large table. This creates a continuous line, a visible backbone that is an explicit representation of company procedures.

The space is lit by a system of indirect lighting positioned in the central part of the tables. Next to the large table a series of cylindrical spaces house meeting rooms, areas for relaxing, and communal areas.

These look like circular pavilions and are lit from inside. In clear contrast with the linear layout of the industrial process, these spaces create a new interior landscape.

Efficient use and control of natural light are obtained by means of closely-woven blinds mounted on rollers and cables outside the building, with a wind-sensitive and light-sensitive mechanism regulating the degree – greater or lesser – of opening.

003/area
general service area

006/lab
005/lab
telecoms test lab
004/lab
raised floor test lab

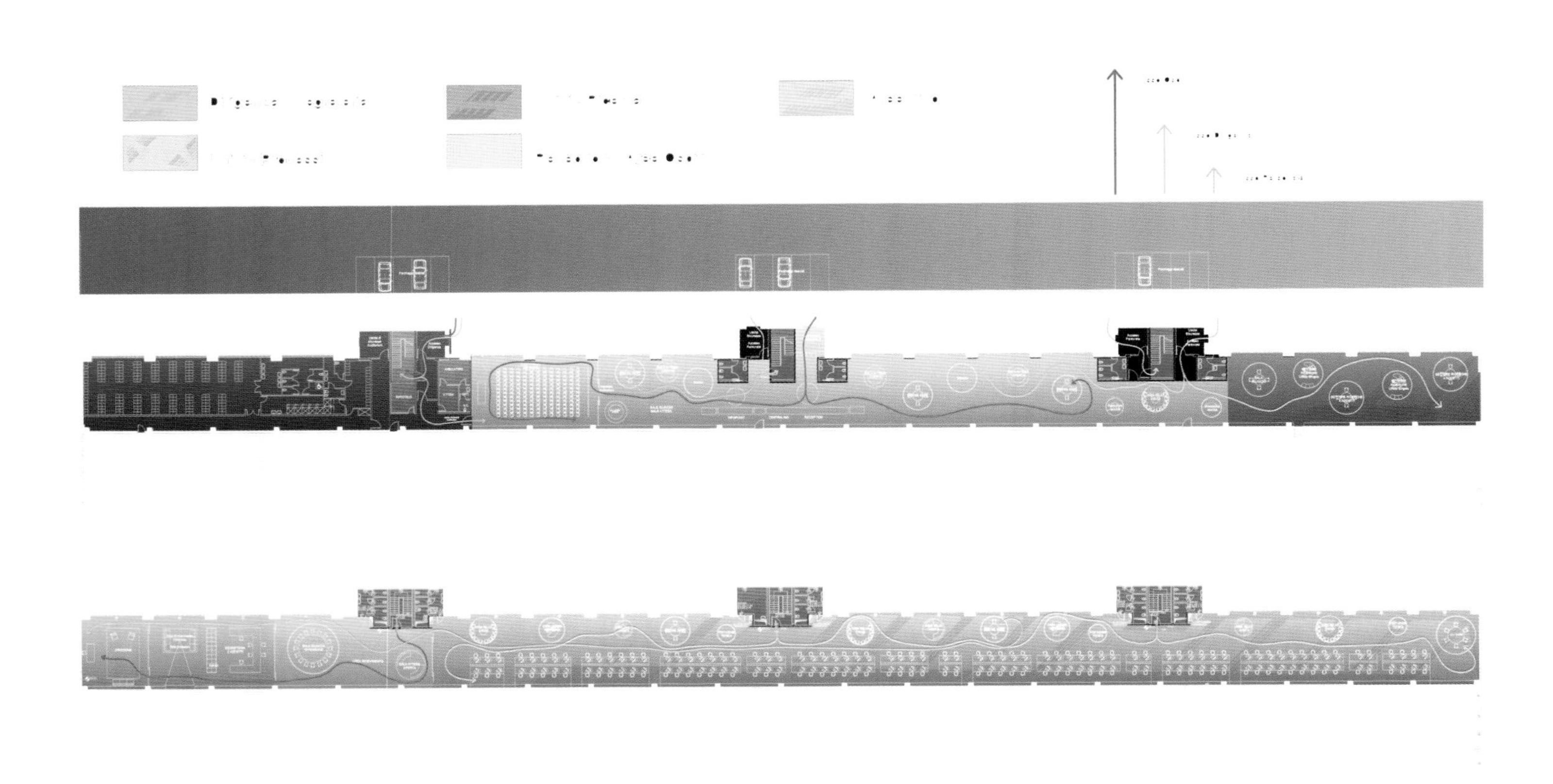

Horus Capital Offices

35

Design Company: Casson Mann

Project Location:Moscow, Russia

Area:2100 m^2

Photographer:Roger Mann

Casson Mann were commissioned to design Horus Capital's new offices in Moscow, with the brief to create a new open-plan working culture for the company. The offices have workstations branching out from a single timber walkway that reaches from the Chairman's office to the board room, finally becoming the board room table. Cellular glass offices for management are embedded in the timber walkway, creating a sense of openness and involvement. The timber runway is cut through by the 'flytower' of the lobby beneath the offices, creating a direct connection between the working spaces and the Museolobby below and allowing glimpses of the mechanism of the flytower. The office's reception area has an interactive reception desk, which reacts and changes with the approach of visitors. Virtual world clocks constantly update to reflect the time in different cities around the world.

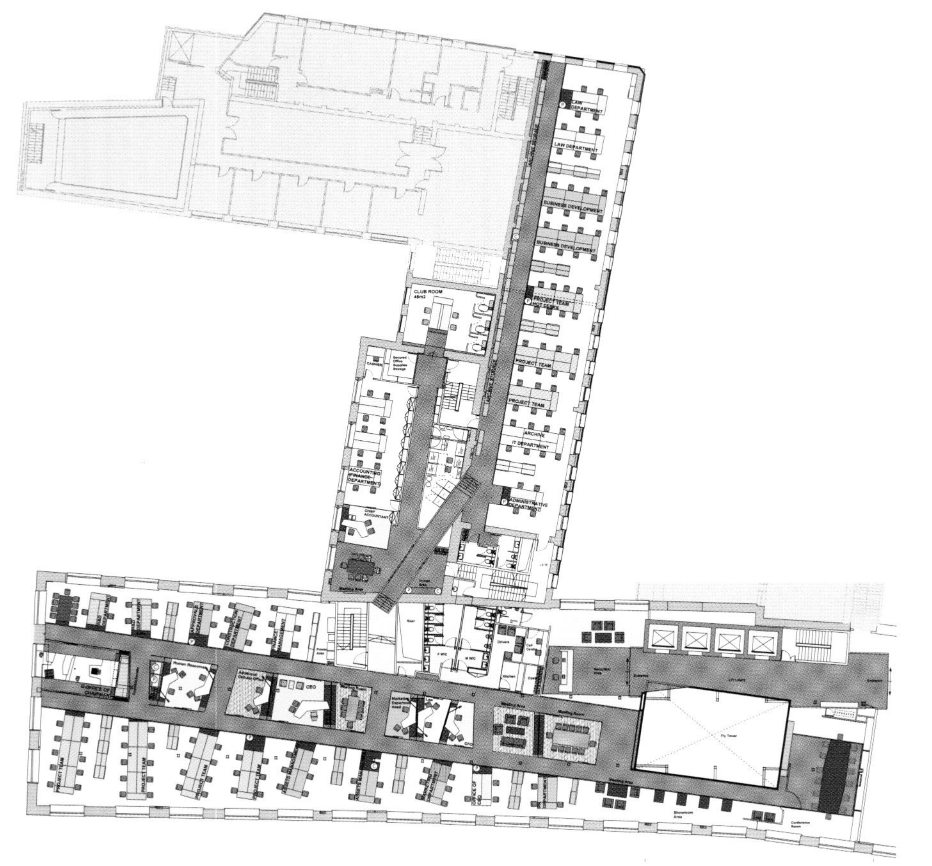

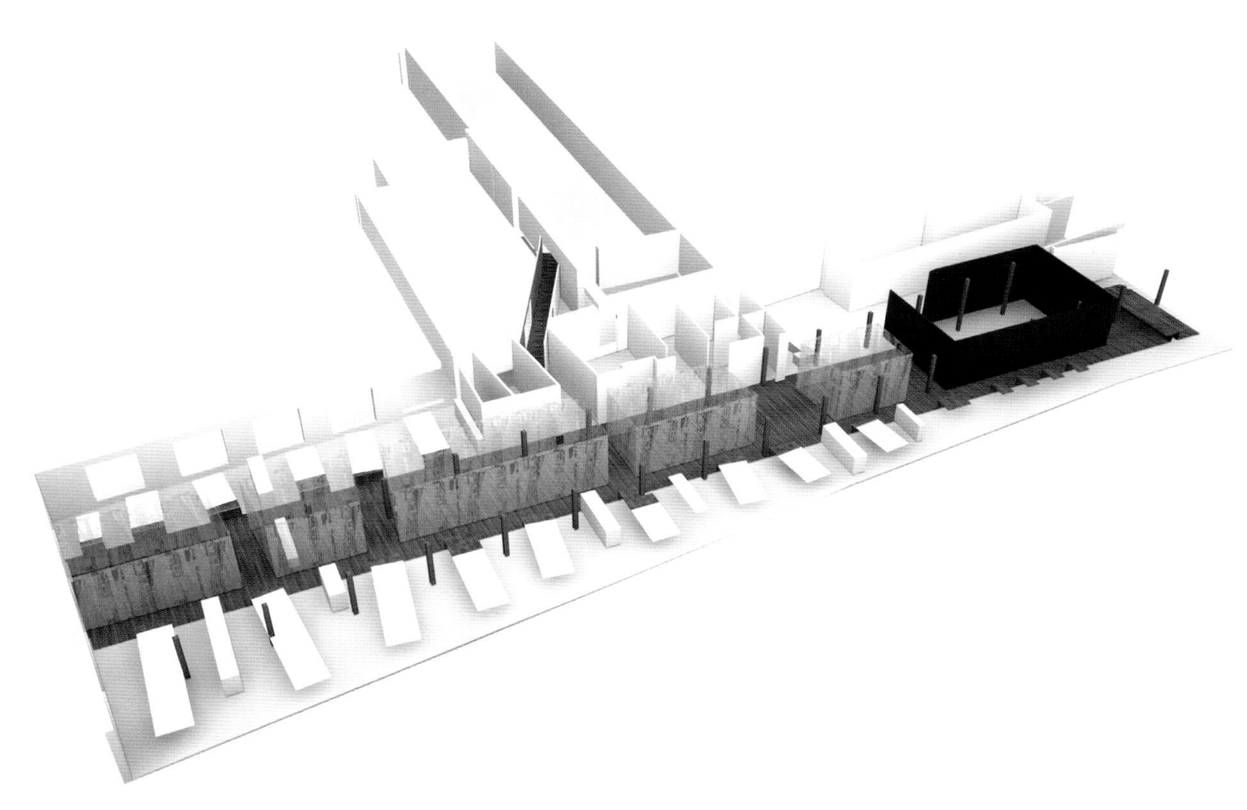

Kao Corporation Head Office

36

Design Company:HaKo Design

Project Location: Nihonbashi Kayabacho Chuou-ku, Tokyo

Area : 1537.5 m^2

Photographer:Nacasa & Partners In. Jun Nakamichi

KAO corporation has started long ago by " KAO SOAP", aiming at fulfilling people's daily need for beauty, cleanliness and health. It is a company which is also always looking ahead and applying the most advanced technology for an ecology orientated business administration. Keeping in mind the company image and policy we decided to design its offices in harmony with those and so we opted for an atmosphere that is soft, mild and smooth, just like a bar of soap. So, the leitmotifs here are white, beautiful, seamless and gently curved surfaces evoking the same warmth and tenderness of one's body skin. We like to think of it as a sort of an organic design. Especially, the dynamic partitions that are placed in the area that widens after passing through the long and narrow corridor leading to the office space, are created in such a way that they provide different impressions depending on the viewing position and viewer's perspective.Also, the surfaces surrounding those partitions are treated with the Japanese historical plastering by traditional plastering craftsmen and earned a smooth whiteness in colour while expressing at the same time company's caring for health.

Furthermore, the waiting room and rest space for visitors are made with actual remnants from the construction works to create an image of forest nature that visually communicates the company's concern for ecology.

Each conference room equipments can be diversified in various ways and this aspect is meant to also stimulate creativity and accelerate motivation during meetings as much as it abounds with originality and serves both functionality and a variety of design.

In order to completely express these design concepts we originally designed all the conference tables for this projects as well.

kao

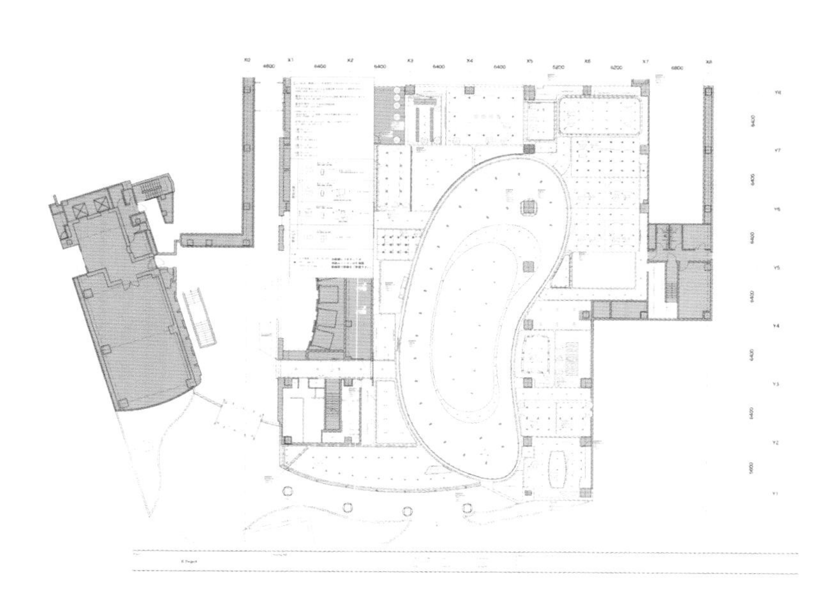

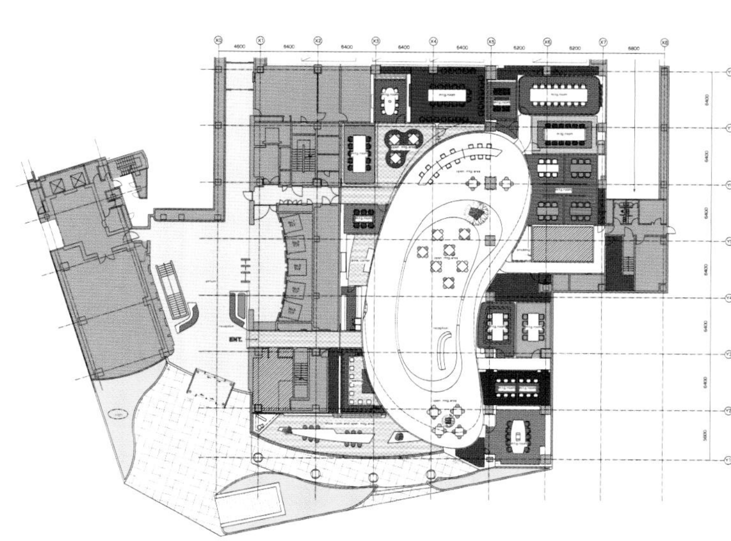

KaO

KaO

Fox Latinamerican Channel Offices

37

Design Company:Alberto Varas & Asociados, arquitectos / Estudio Angelica Campi
Project Location: 5517 Honduras Street, Buenos Aires City, Argentina
Area: 3200 m^2
Photographer:Gustavo Sosa Pinilla

A highly technological building had to be built in a plot occupied by a concrete structure which belonged to an old warehouse. The structure, of a high bearing capacity, covered the whole plot and had no communication with the exterior, because of its use as a warehouse, but it had to be maintained, as much as possible, for financial reasons.

On the other hand , the new building demands were: to get natural light for working areas and to distribute the complex supply net of informatics, telephonic, and satellite services, air conditioning, and energy required by the new functions, taking this net into the working areas, offices and meeting rooms that were part of the different areas of the layout.

In this way the problem became a game of interior perforations to get natural light in the interior space and to generate a spatiality according to the creative labours of each of the varied sectors, which means, a system of differentiated and interconnected spaces, which where inserted in the in- between of the only two facades of the building able to connect with the urban space. One of them, the backward facade, was the one in which a four floors height vertical garden was generated to deliver views and light to each floor of the building. The other, the street facade, consisting of a double facade, one operable and the other fixed which regulates the entrance of light and reflects the building's technological character.

Both facades reveal in a small scale the demand that exists in actual high-tech buildings, consisting in compensating the hipertechnological scenario with some way of natural presence.

The glass façade looking towards the street, a ventilated facade of frosted and tempered glass is made up of glass panels where the grinding is integrated as a series of dots with different density according to the front sector they cover, until they vanish away including the supporting glass, this, in coincidence with the areas where windows are open and predominate the interior views to the outside.

So the facade appears as "pixelated" by the different effects of the grinding, referring to the "visual atom" which is the prototypic material handled inside the building: the digital image.

During the evening a lightening system with leds and projections on a screen on top of the facade, allows it to get totally coloured being neutral during the day, This converts the building in some kind of spectacle over the street, particularly in a neighbourhood where nightlife and the media companies have settled during the last years.

Passing through the frontal façade to the building's back is a theoretic promenade which leads from the public world of technology, the sophisticated leds light and the data projections over the façade into the quietness of the garden and the silence of the internal working areas where people can briefly and hopefully imagine to escape from the intriguing and seductive digital machine inside of which they work.

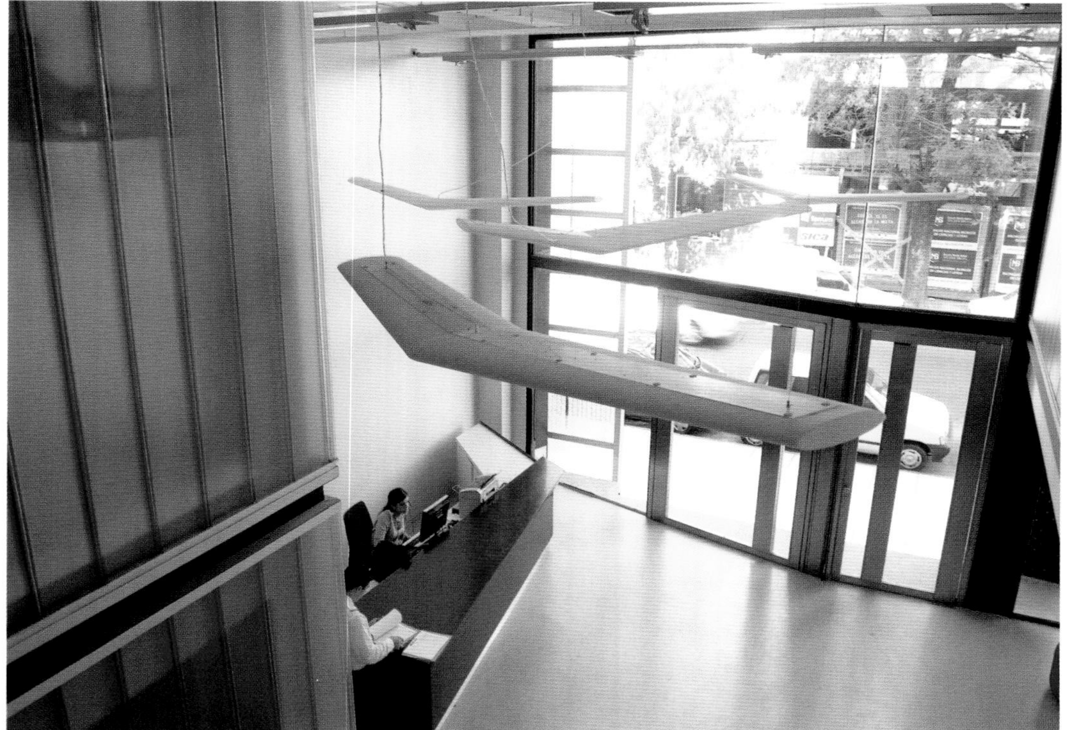

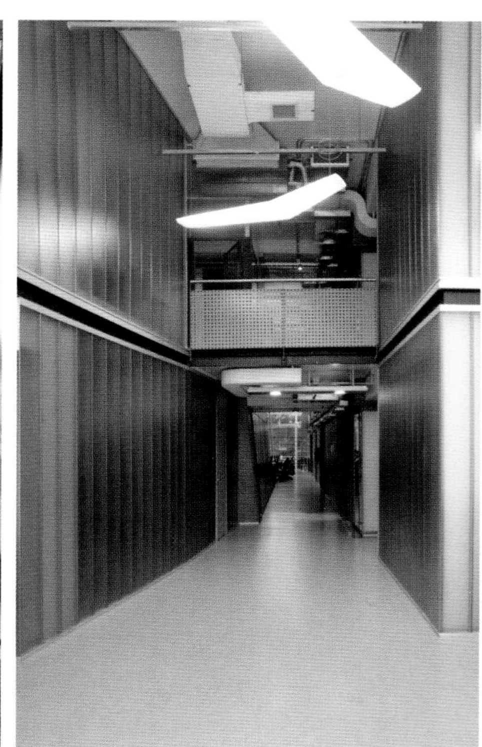

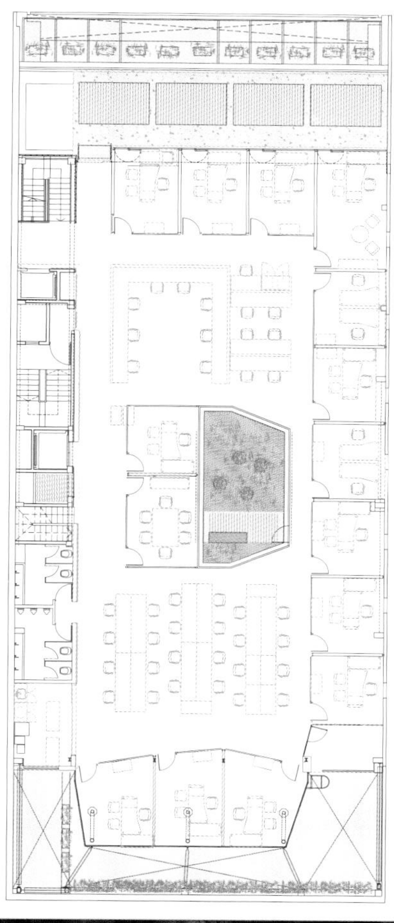

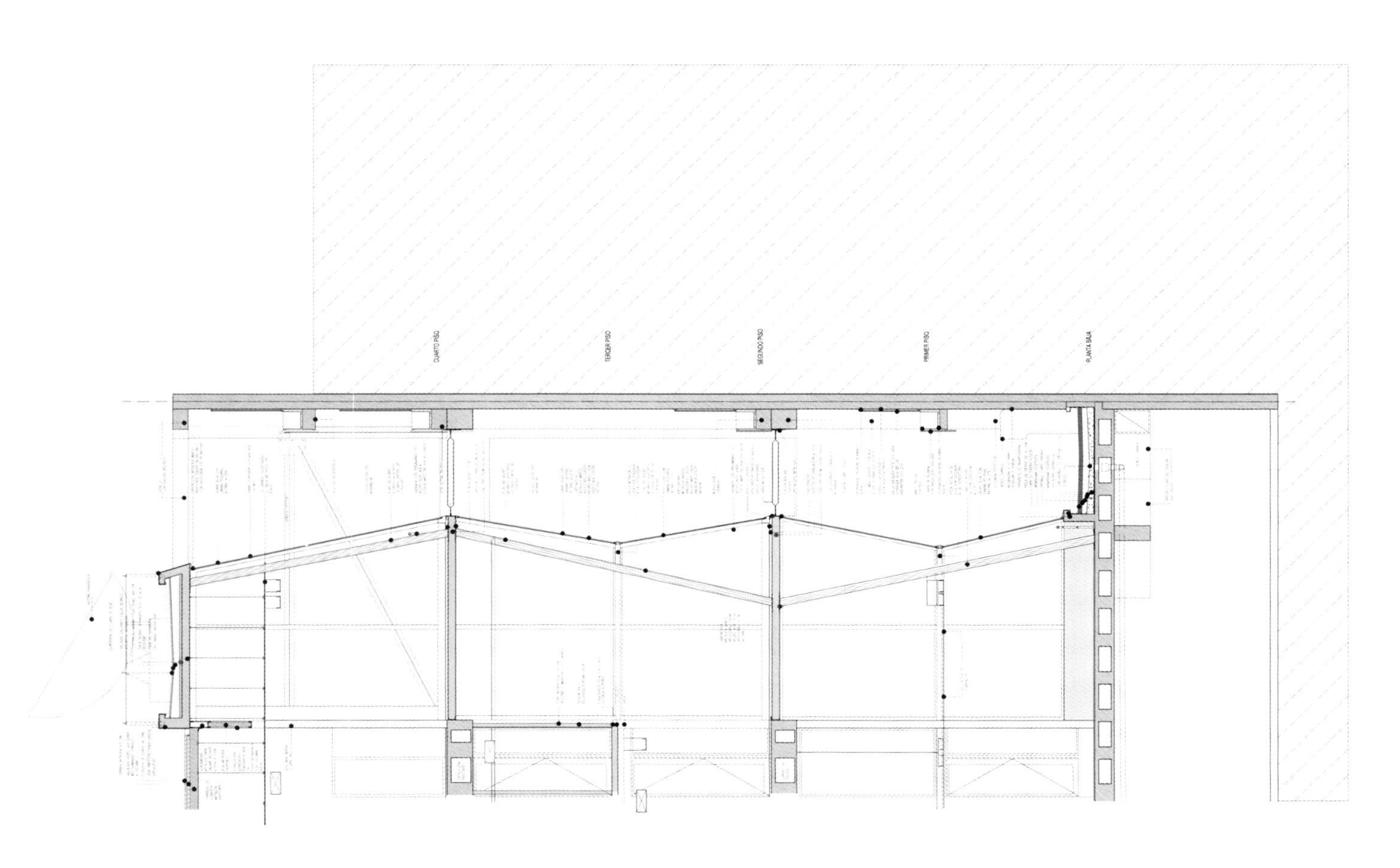

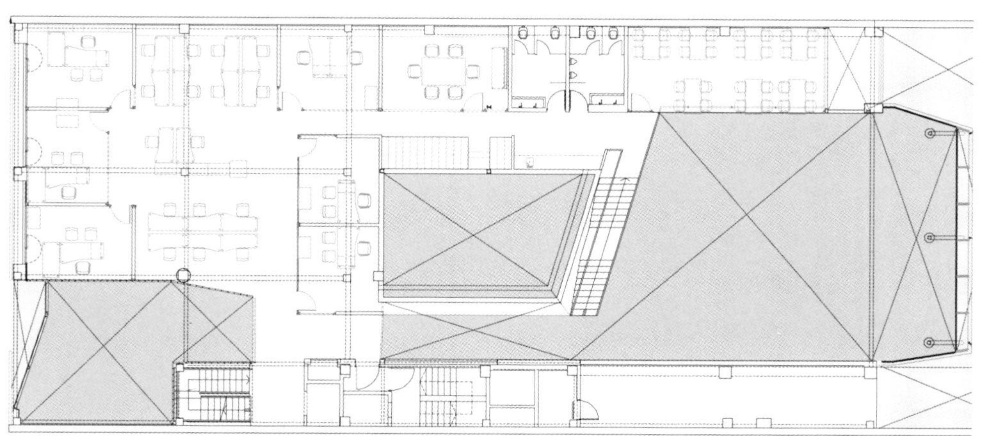

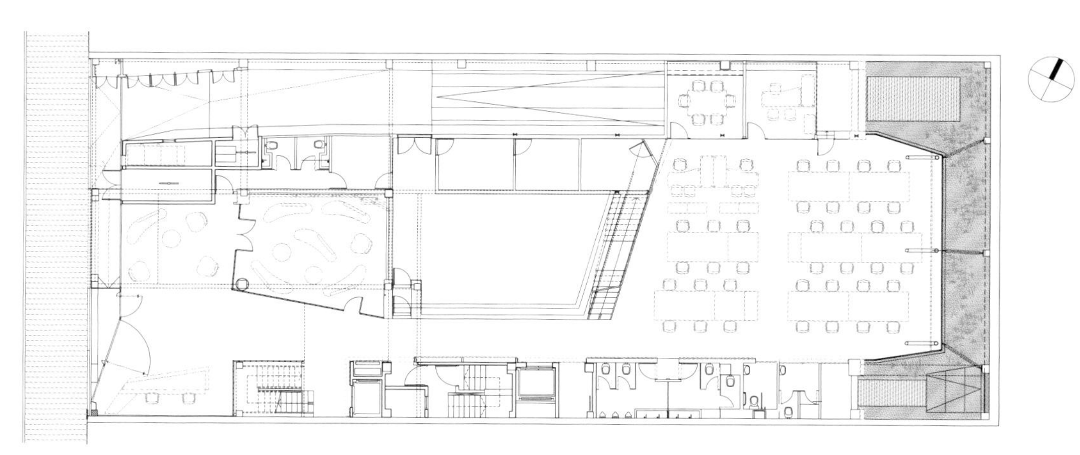

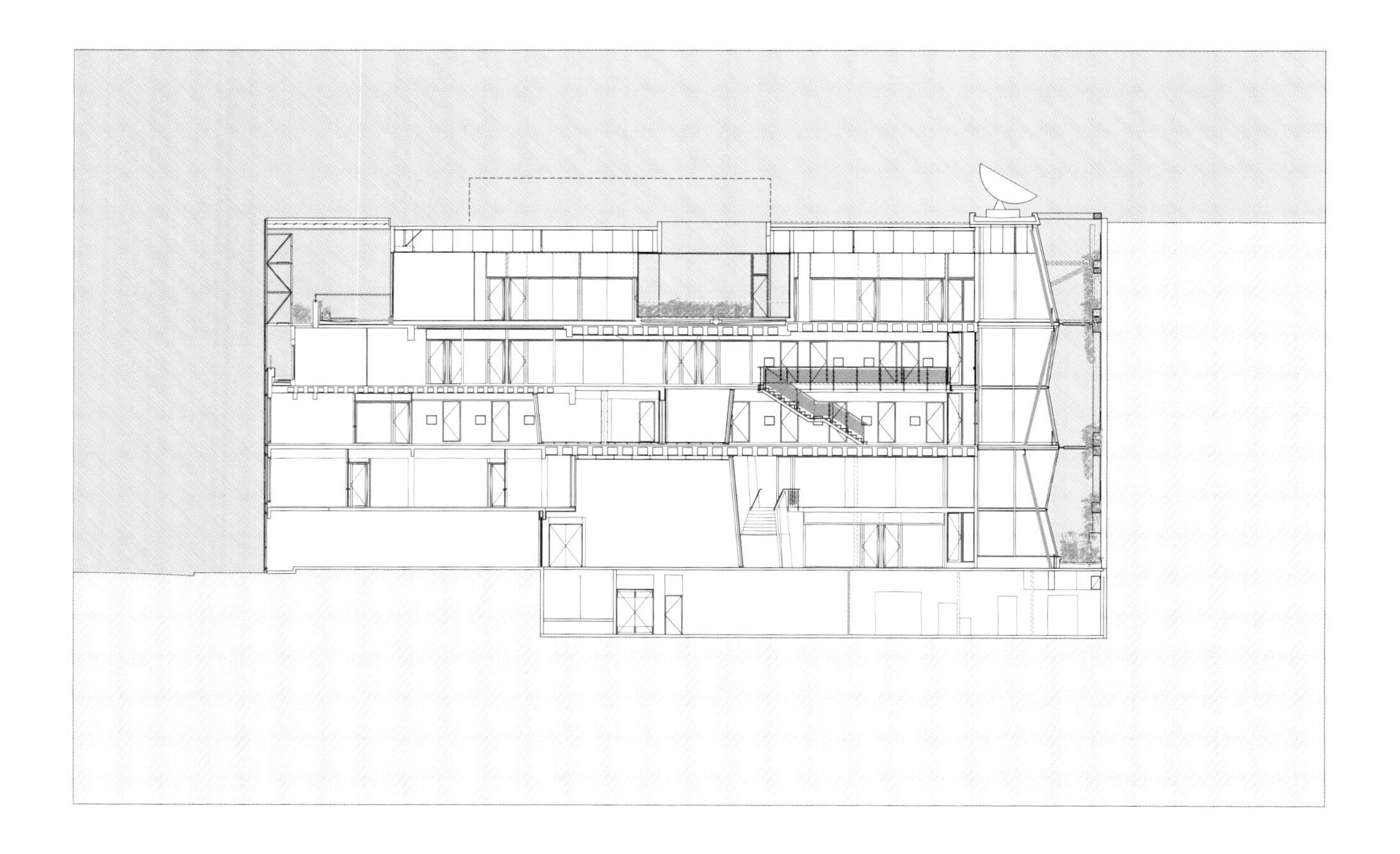

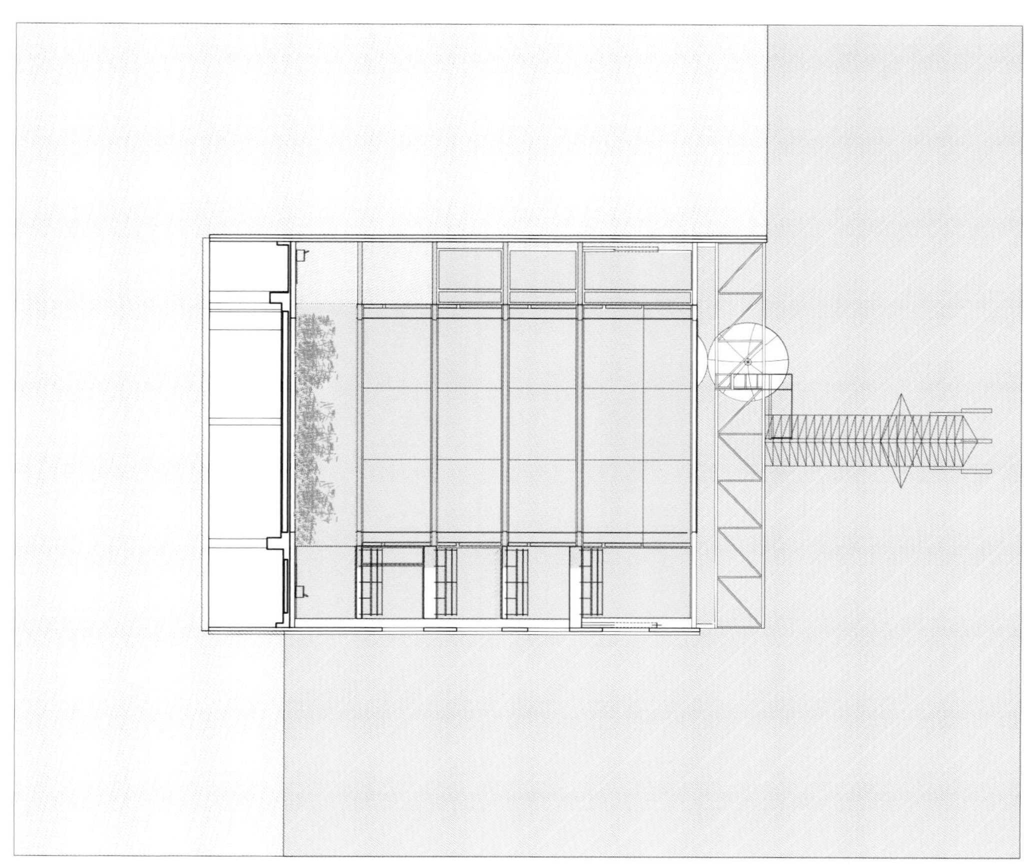

Design Offices

38

Design Company: SPG Architects

Project Location: New York City, NY, USA

Area: 241.54 m^2

Photographer: SPG Architects

SPG Architects designed this space as a flexible open-planned design studio. Vertical planes of eco-resin and acrylic transmit and reflect light and act as foils to the horizontal work surfaces and floating storage units. The sense of openness is created by the continuous ceilings, which overlap contiguous spaces. The ceilings further define the character of the studio by acting as the means to light the space, either through coves in discrete areas or, as in the main linear studio, as a reflector for the custom linear pendant up-light fixtures. The entire office is washed with natural light by a wall of windows that unifies this interior with the city's energy yet maintains a serene effect through its monochromatic palette of grays, creams and whites.

COLOR
1
2
3
4
5

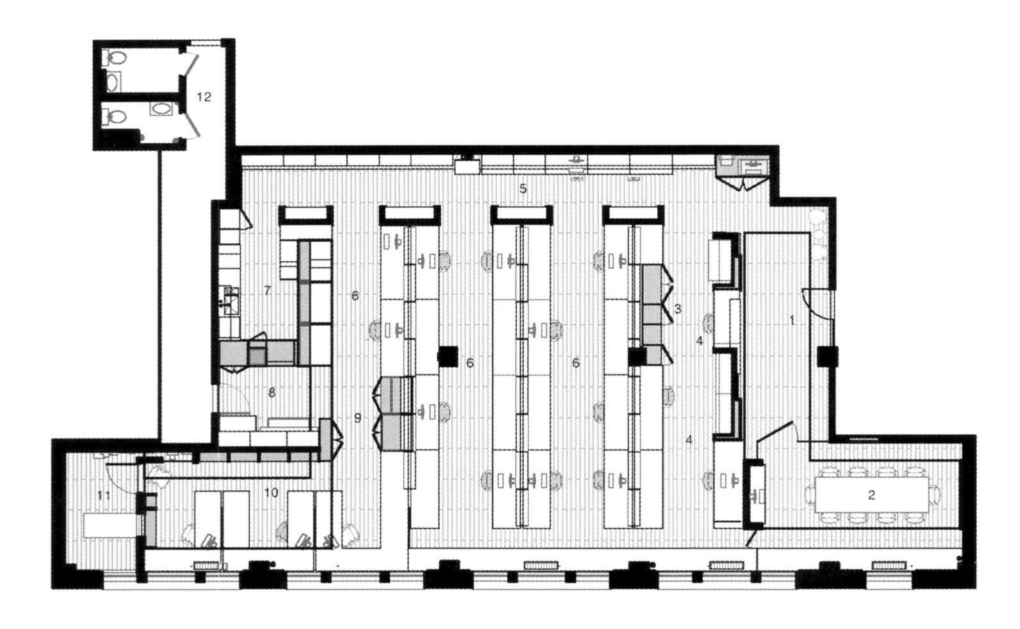

5' 10' 20'

1. Reception
2. Conference Room
3. Supplies Closets
4. Receptionist / PR
5. Library
6. Studio Space
7. Kitchen
8. Printing Services
9. Coat Closet
10. Partner's Wing
11. Mechanical Room
12. Bathrooms

FLOOR PLAN

1-10design Kyoto Office

39

Design Company:TORAFU ARCHITECTS

Project Location: Karasuma Kyoto

Area : 460 m^2

Photographer:Daici Ano

The offices of Kyoto-based web production company, 1-10design (one to ten design), comprise a single 460 m^2 versatile area featuring work spaces, a meeting room, a gallery and a traditional Japanese room serving as a resting area.

These offices are not only fit for business, but for a variety of purposes as well. We proposed a spacious floor finely partitioned by wooden frames, allowing rooms of various sizes to flow into each other. The framework can also serve as shelves on which we can find office furnishings alongside personal belongings. The frames keep functional spaces separated as the rooms' appearance change every time one passes through them, and yet, they offer a peering view that gives the whole floor a sense of continuity.

The main work space straddles many rooms but a long table running across the center connects them together. Adjacent to this work space is a meeting room, a laboratory, the president's office, a traditional Japanese room and a gallery. Also, the wooden frames offer flexible storage combinations and a variety of usages by turning into book shelves, stowage space or even a bench according to location.

This bare skeleton can become a wall as much as a piece of furniture but really serves as a background prompting miscellaneous interactions by keeping the ability to accommodate future changes with flexibility.

The office allows one to change the scene in front of us according to our mood while preserving the sense of unity of a single room.

INTERACTIVE STUDIO

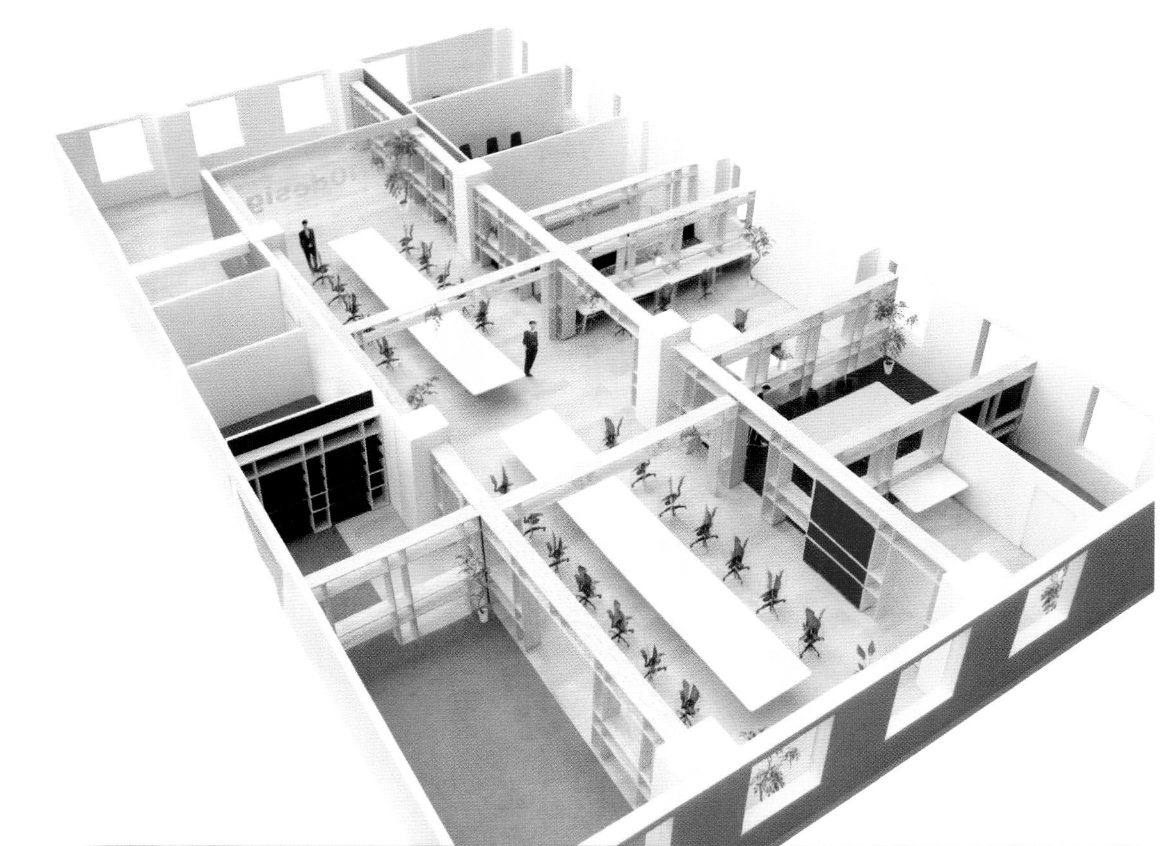

MEETING ROOM
CHERRY

KAYAC Ebisu Office

40

Design Company:TORAFU ARCHITECTS

Project Location: Ebisu Tokyo

Area : 249.9 m^2

Photographer:Daici Ano

KAYAC is an IT business that is also into the "Fun-loving Business", so we thought of a flexible and adaptable space for their new Tokyo branch office where employees could enjoy visualizing their company's performance.

Project teams need to change constantly and fluidly and so does the desk layout in the office. We thus proposed a design that offers each member a private desk that can be reshuffled at will into a temporary configuration and accommodate future additions to the team.

We also fitted wheels on regular desks, shelves, partitions and planters just like strapping roller-skates on to make them go mobile and create a flexible office space unhindered by fixed furniture. Colored stripes found at the edges of the carpeting gently indicate the zoning around each area.

The side facing the entrance features windows and a wall where each month's corporate performance can be visualized. Colored balls on a mesh mark the sales made so employees themselves can update the ledger daily. This colorful graph is part of an ensemble that contributes to add color to the office space.

By giving the office a flexible configuration that does not overly monopolize the entire space and by keeping corporate performance results in the field of vision at all times, we sought to fit the company's new working space where they themselves propose to others a new working style.

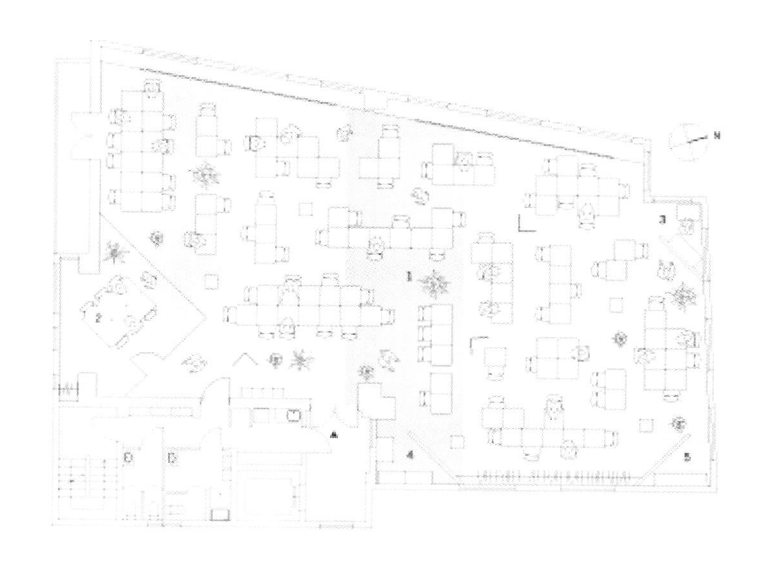

KAYAC

7
8
10
11
50

So Architecture office

Design Company: SO Architecture - Shachar Lulv & Oded Rozenkier

Project Location: Shaar Haamkim, Israel

Area:72 m^2

Photographer: Asaf Oren

Nestled in the heart of Kibbutz Shaar Haamkim, So Architecture's office features a cohesive rectangular office space that appears to emerge from the hill top upon which it sits. Extruding the landscape into smooth hill-like contours, public access is allowed through a dense array of foliage, which gradually opens out into a clearing to accommodate for the communal gatherings of the kibbutz. To a passerby, the office itself seems to blend into the existing building, which used to be an old sports hall. Visually, too, each tenant in the office enjoys the impression of a workspace closely attributed to the surrounding landscape due, in part, to the large panoramic window that provides a spectacular view of Tivon's hills. Arranged linearly along the expanse of glass overlooking Tivon, the working stations fulfill the purpose of not only providing for the needs of the architects, but also help to establish the necessary working environment for dynamic communication. As a result of the long oak table which unifies the workstations of all architects within the office, interaction is made effortless for each has the means to assist and view each other's work. Separating the workspace from the smaller conference table located in the far parameters of the office is a large glazed frame illuminated by a laser-cut light in flush with the ceiling. However, since this division is not made palpable by any physical barriers, the meeting space and adjacent workspaces of the architects are seamlessly joined together. By committing the work of the office to the collaborative design process with the client, So Architecture places precedence not only to the artistic merits of the final product, but also the process involved in the creation of architecture. Hence, So Architecture imbues their location with the essential spirit of the community in which they work and through an inclusive and collaborative design process, productively redefine the working environment.

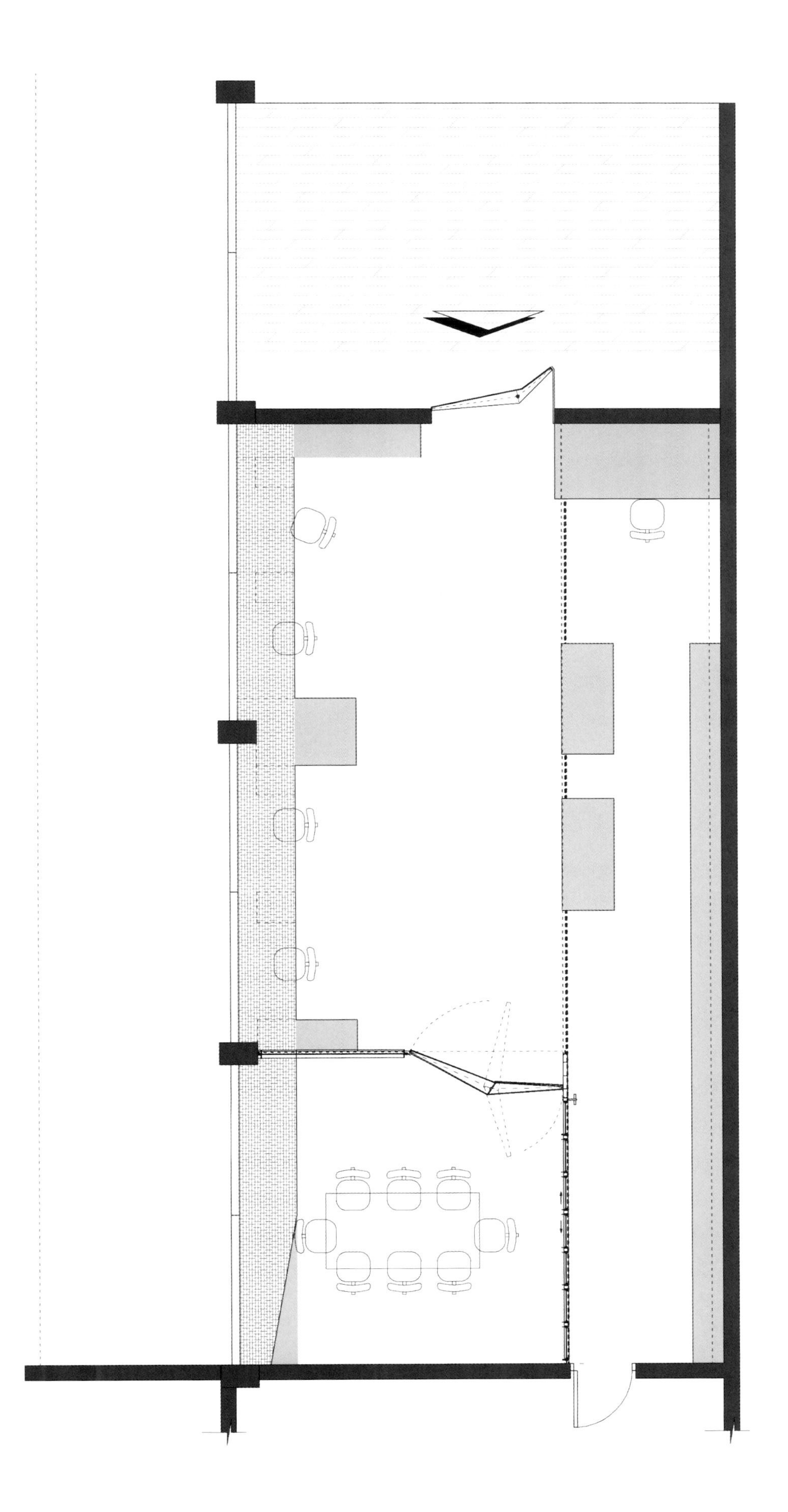

ARCHITECTURE

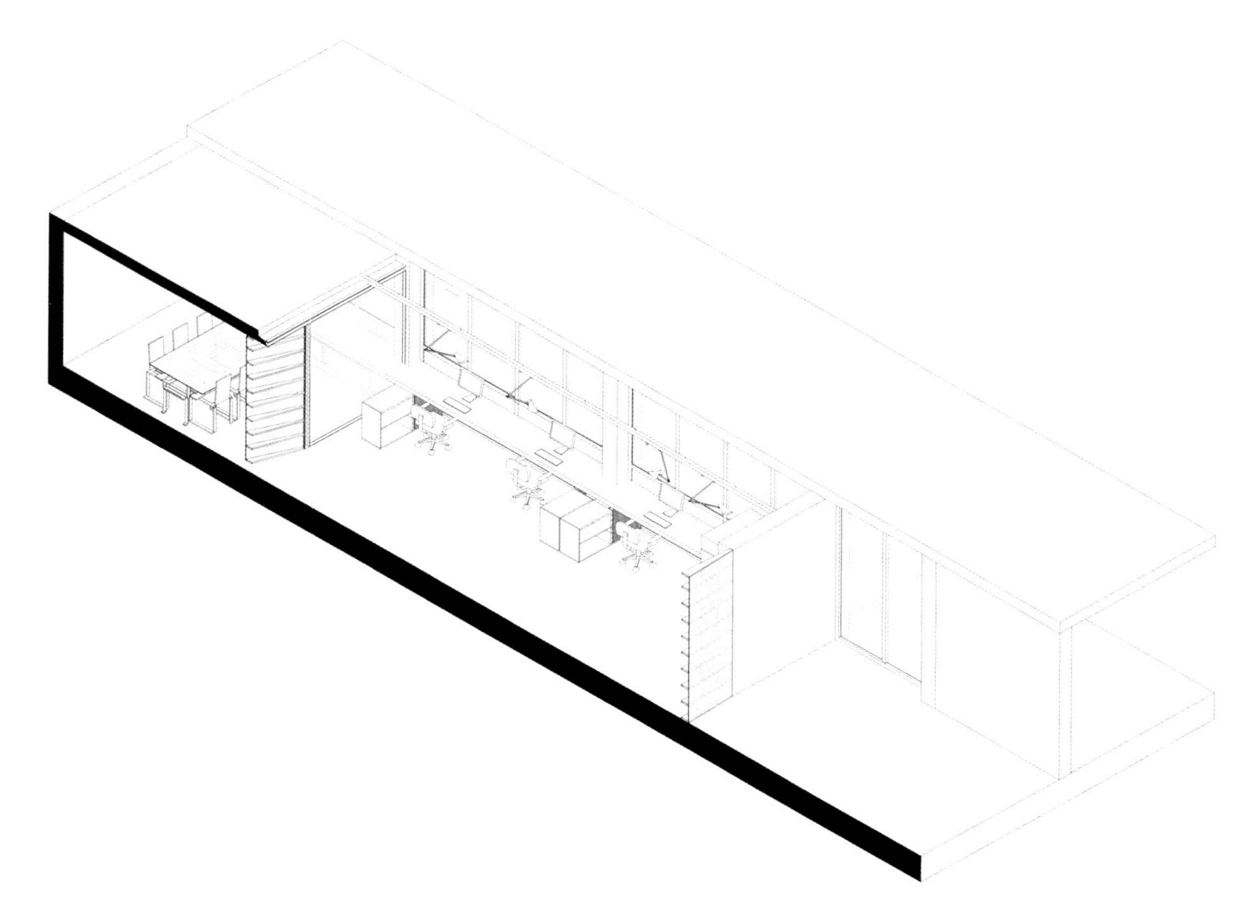

RETAIL

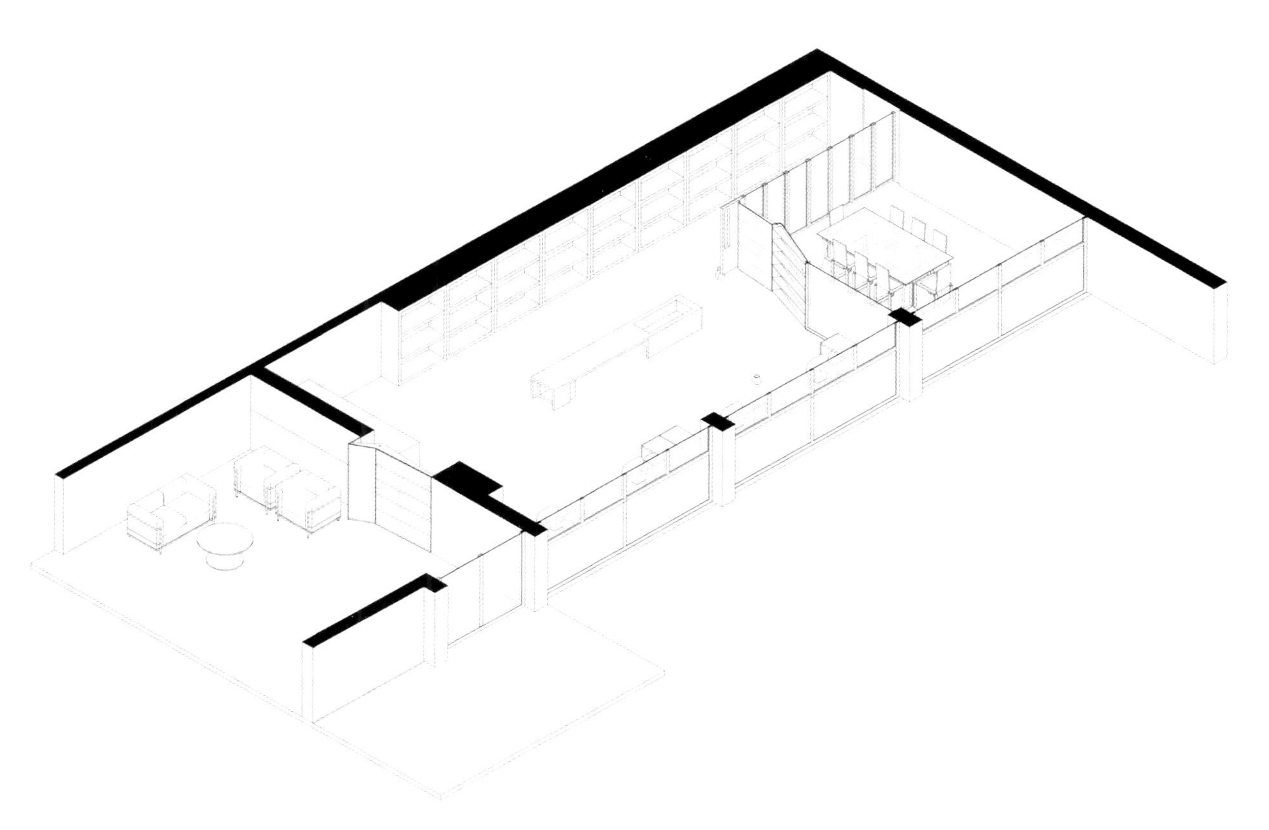

Ericsson Office(Innovation Room)

42

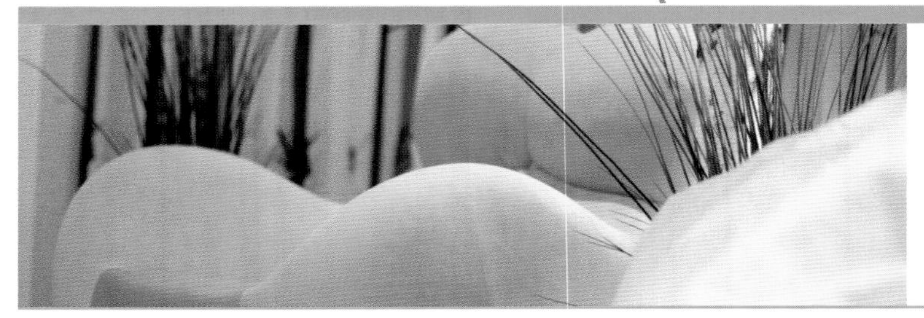

Design Company: DPWT Design Ltd

Project Location:Beijing,China

Area :50 m^2

Photographer : Chen Bin

This project is a renovation project of Ericsson Zhaowei office. It transforms the original meeting room into an "Innovation room", a room that inspires employees' innovative thought.

We incorporate the concept of green space into the design. Seeing the bamboo forest and emulational silver birch on the curtains, you seem to smell the fresh air in nature. The suspended ceiling bulbs are like fireflies flying in the thicket. The comfortable and interesting sofa is made over from a small boat. Through the two "eyes" on the entering door, passersby can peep at the special room.

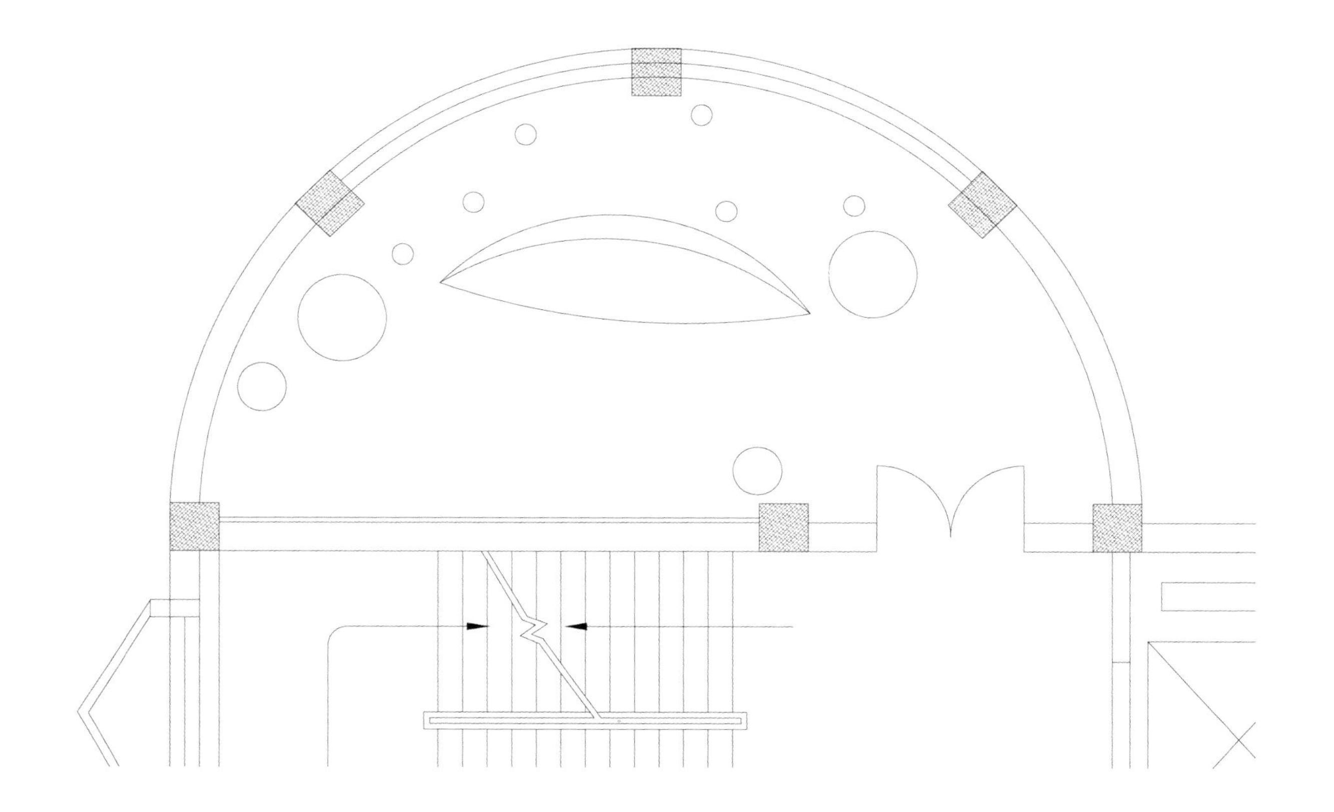

Headquarters Office of Beijing Wintop

Design Company: DPWT Design Ltd

Project Location:Beijing,China

Area :4000 m^2

Photographer: Feng Zhiguang

The headquarters office of Beijing Wintop is on the fourth and fifth floors of the building. On the fourth floor there is a reception counter, major meeting room, opening office, manager office and so on. The fifth floor is for such senior office areas as the independent board room, study room and accounting office. The holistic design has both strong modern atmosphere and abundance of Chinese style. The superiority of the room space and the professional choice of materials have exhibited the qualities and characteristics of Wintop Estate to the point. The design of lighting and color makes the modern atmosphere prominent in the space. It is the lobby design that satisfies us most. There are added spotlights on the LOGO wall with the length of over 5 meters, which has highlighted the visual key point. Beside each footstep of arc-shaped stairs that connect the ground floor and interlayer, there is an LED light. There is a round light box on the false ceiling of the glazed mini meeting room which symbolizes the Chinese lantern. It is the focal point of the whole design and has promoted the company image, intensified the sense of space and made it stand out from all the designs.

创造最具价值的生活空间
To create the most valuable life space
万通地产
VANTONE

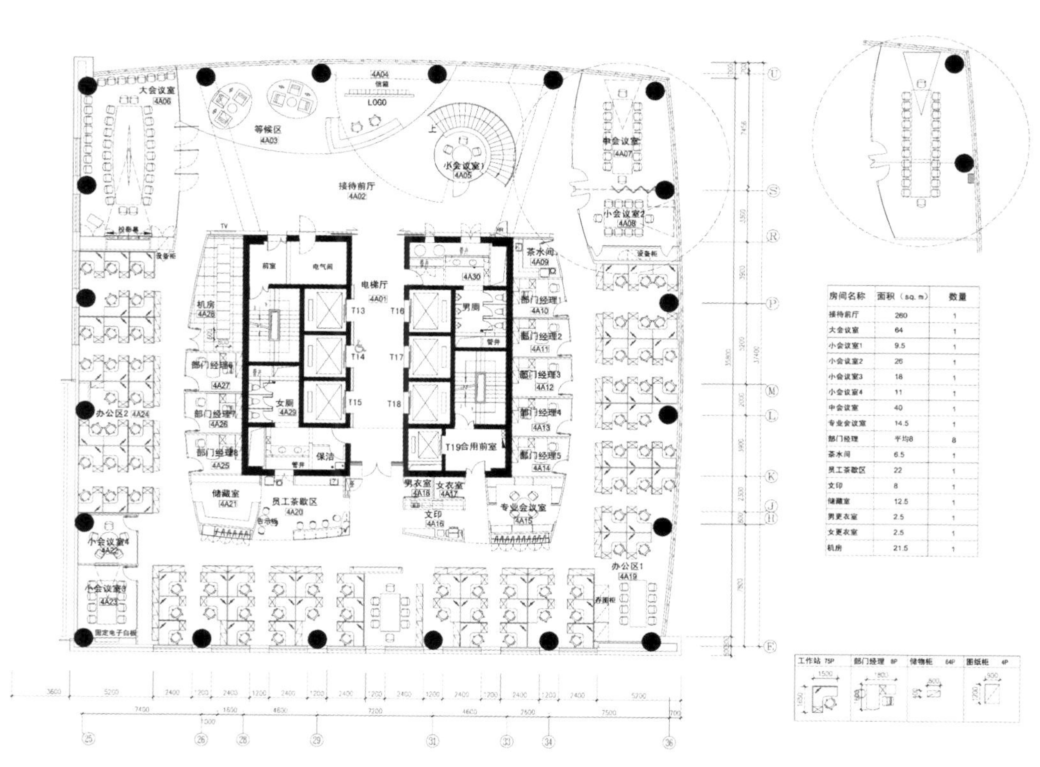

房间名称	面积（sq. m）	数量
接待前厅	260	1
大会议室	64	1
小会议室1	9.5	1
小会议室2	26	1
小会议室3	18	1
小会议室4	11	1
中会议室	40	1
专业会议室	14.5	1
部门经理	平均8	8
茶水间	6.5	1
员工茶歇区	22	1
文印	8	1
储藏室	12.5	1
男更衣室	2.5	1
女更衣室	2.5	1
机房	21.5	1

Dell Tokyo Office and Show Room

44

Design Company: Hiroshi Shirasaki, PLANET DESIGN AND CONSULTING

Project Location: Tokyo Japan

Area: 1000 m^2

Photographer: Nacasa & Partners

Tokyo and Osaka offices are designed to represent DELL's new brand images.
The direction of the enterprise that DELL is driving for is represented by advanced materials and clean details. How DELL's technology can arrange information beautifully and speedily among the modern society in which information, including internet, mobile phones, etc., is flying around......such images are represented in the space. Entering the reception, Office zones are also designed with motif of white, silver and CI color blue, and arranged by advanced images.

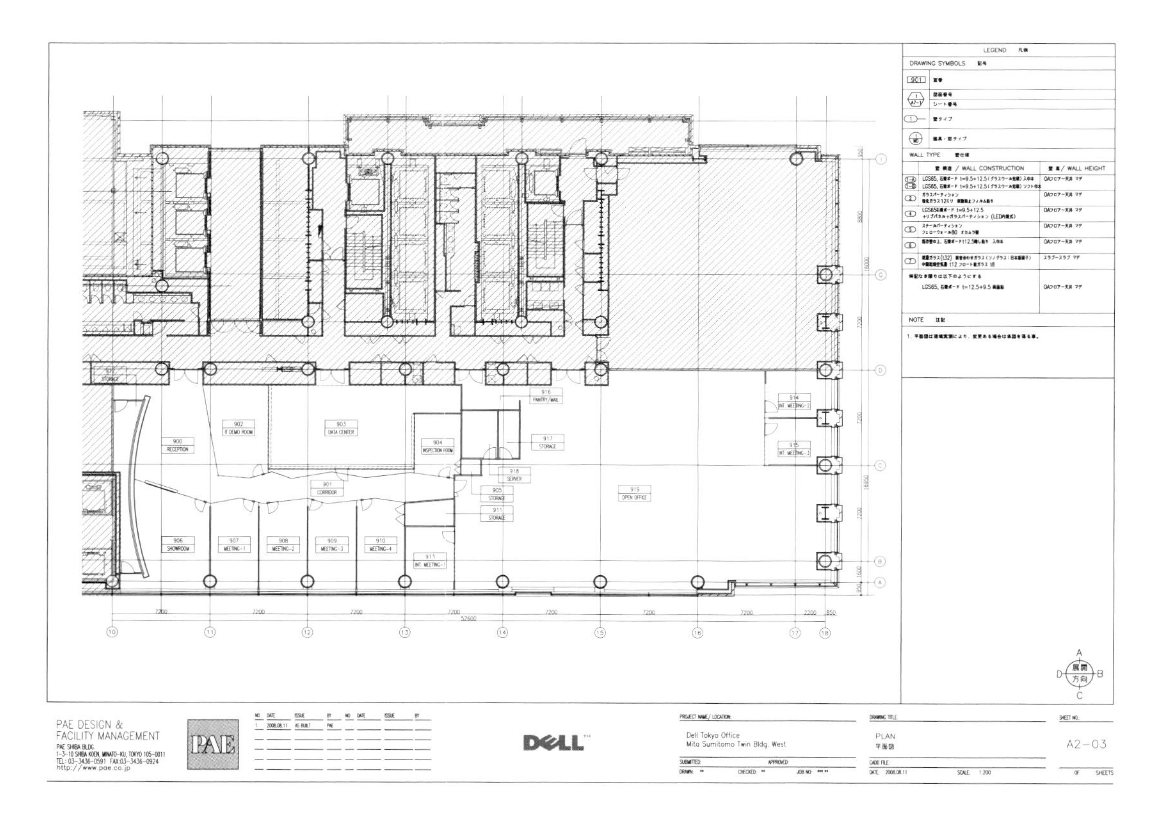

LEGEND 凡例
DRAWING SYMBOLS
WALL TYPE
WALL CONSTRUCTION
WALL HEIGHT
NOTE
RECEPTION
IT DEMO ROOM
DATA CENTER
INSPECTION ROOM
CORRIDOR
SHOWROOM
MEETING-1
MEETING-2
MEETING-3
MEETING-4
INT. MEETING-1
PANTRY/MAIL
STORAGE
SERVER
OPEN OFFICE
52600
PAE DESIGN &
FACILITY MANAGEMENT
PAE SHIBA BLDG.
1-3-10 SHIBA KOEN, MINATO-KU, TOKYO 105-0011
TEL: 03-3436-0591 FAX:03-3436-0924
http://www.pae.co.jp
PAE
2008.08.11
AS BUILT
DELL
PROJECT NAME/ LOCATION:
Dell Tokyo Office
Mita Sumitomo Twin Bldg. West
DRAWING TITLE
PLAN
平面図
SHEET NO.
A2-03
SCALE 1:200

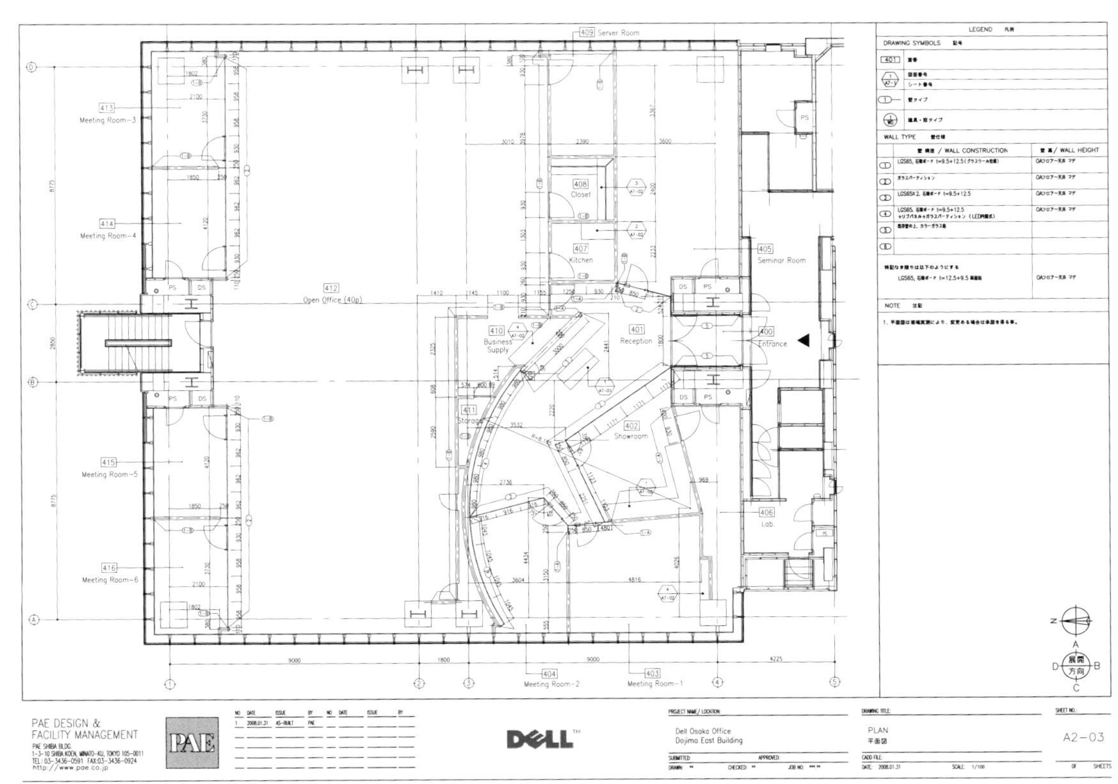
409 Server Room
413 Meeting Room-3
414 Meeting Room-4
415 Meeting Room-5
416 Meeting Room-6
412 Open Office (40p)
408 Closet
407 Kitchen
405 Seminar Room
410 Business Supply
401 Reception
400 Entrance
411 Storage
402 Showroom
406 Lab
404 Meeting Room-2
403 Meeting Room-1
LEGEND
DRAWING SYMBOLS
WALL TYPE
NOTE
PAE DESIGN &
FACILITY MANAGEMENT
PAE SHIBA BLDG.
1-3-10 SHIBA KOEN, MINATO-KU, TOKYO 105-0011
TEL:03-3436-0591 FAX:03-3436-0924
http://www.pae.co.jp
PAE
1 2008.01.31 AS-BUILT PAE
DELL
Dell Osaka Office
Dojima East Building
PLAN
平面図
A2-03
DATE: 2008.01.31
SCALE: 1/100

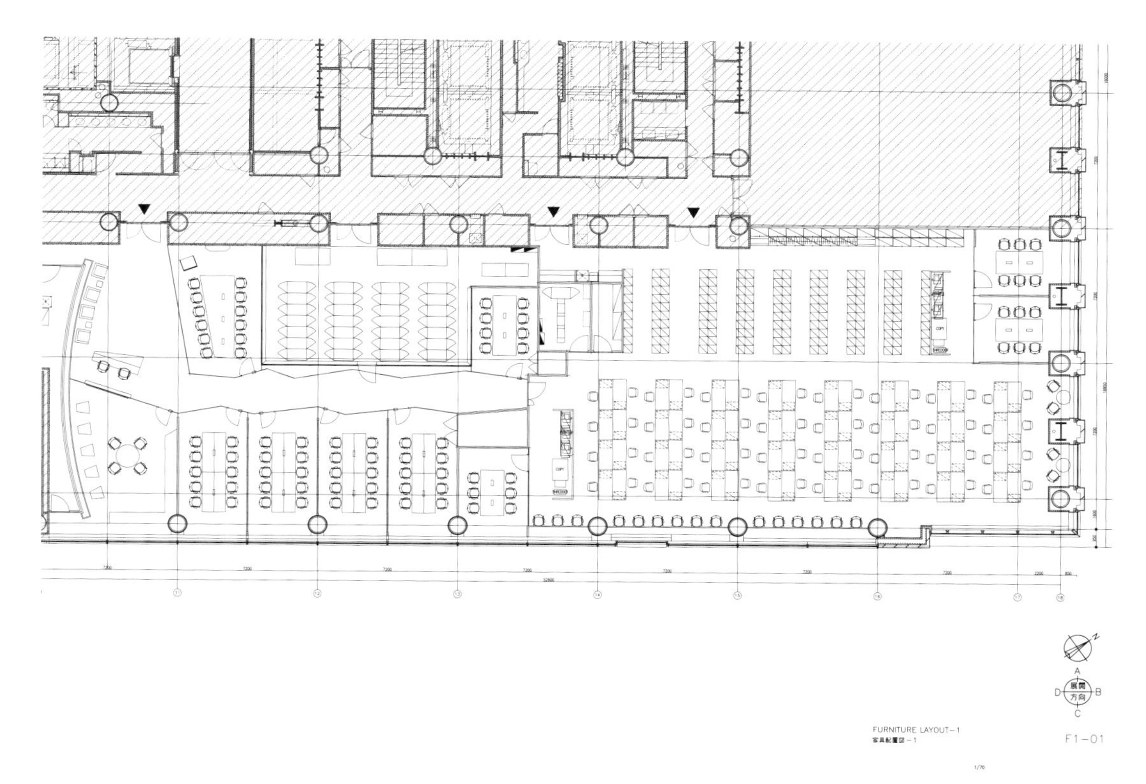

展開
方向
FURNITURE LAYOUT-1
家具配置図-1
F1-01
1/70

M&C SAATCHI Advertising Agency

45

Design Company:Dale Jones-Evans Pty Ltd Architecture

Project Location: Sydney Australia

Area: 2918 m²

Photographer:John Gollings

Situated opposite the Botanical Gardens within Transport House, a Heritage listed Art Deco sandstone building, the new fit out occupies 4 levels of this beautiful building. As the agency had out-grown the previous office, the main idea was to bring together and integrate all the disparate parts of this large 300 persons organisation. And to do it in such a way that everyone remained connected. The grand hall therefore acted as the central creative engine of activity and exchange.

The existing building had many impressive Art Deco elements, especially the dual street heritage lobby entries and old Registration Hall. The main 2 storey hall was therefore designed to house the working hub of the agency and provide a powerful public domain atmosphere. The main hall integrated both the Creative and Accounts departments, separated yet also gathered everyone around a huge long communal coffee table placed central in the hall. New works were kept to a minimum and designed to reinforce the halls former layout. Joinery counters of rich lacquered wood were placed to reinforce the halls former glory.

The main lobbies were restored and juxtaposed with a contemporary café and bar. The café bar acted as the public interface of the organisation and a place to view wall projected videos 'facebook' style portraits of staff and their campaigns. A large serpentine shaped plywood table arcs across the café floor providing numerous network points for staff and visitors to both informally recreate and work. The space is back dropped to a black floor and café counter with fresh coffee and food – itself wrapped in the former heritage rooms white painted walls and ceiling.

Formal and informal meeting rooms are sprinkled throughout the interior with the larger 20 to 30 seat boardrooms flanking the street scape and park views. Upper level floors and the grand halls mezzanine house other departments and the directors headquarters.

The basement floors were treated as warehouse type spaces and simply painted white to ensure plenty of light and housed the buildings ceiling expressed services. They were dotted with workstation and furniture choices containing splashes of colour.

M&CSAATCHI

PHILLIP STREET ENTRANCE

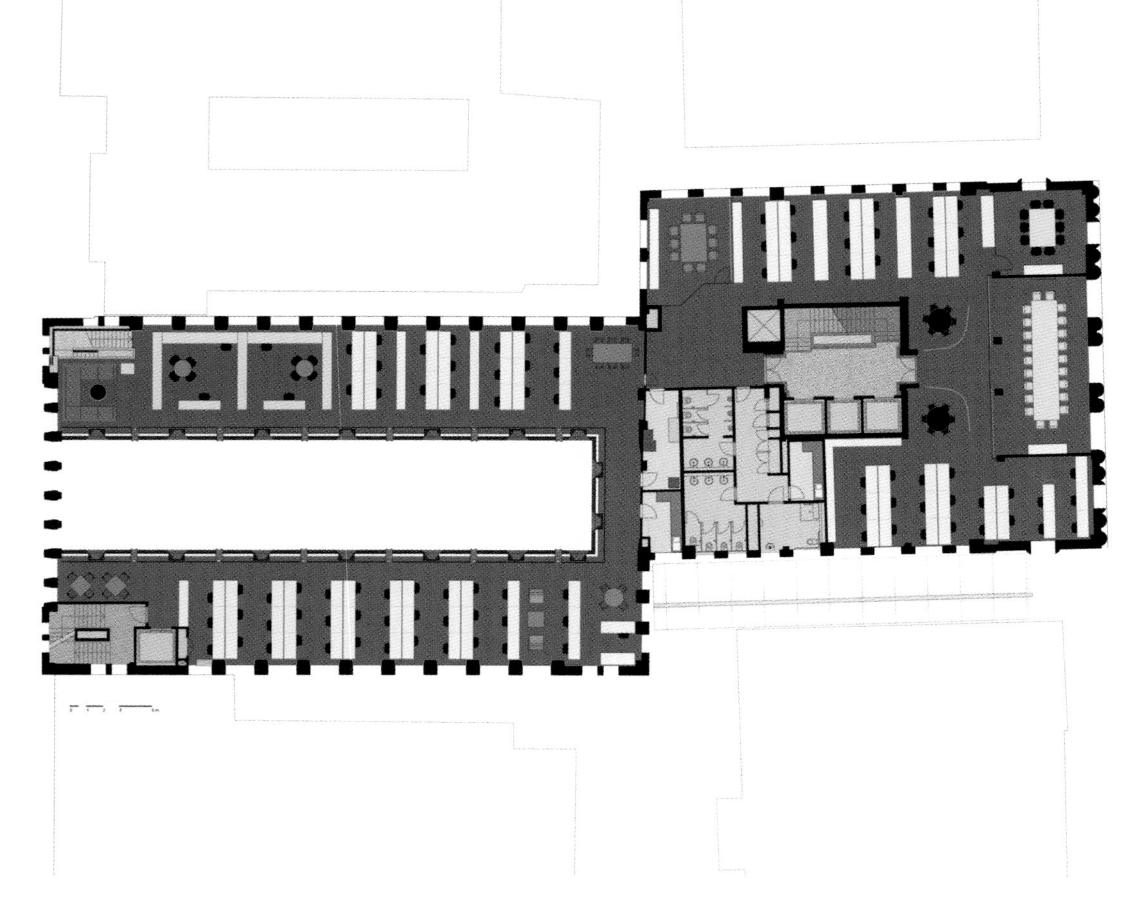

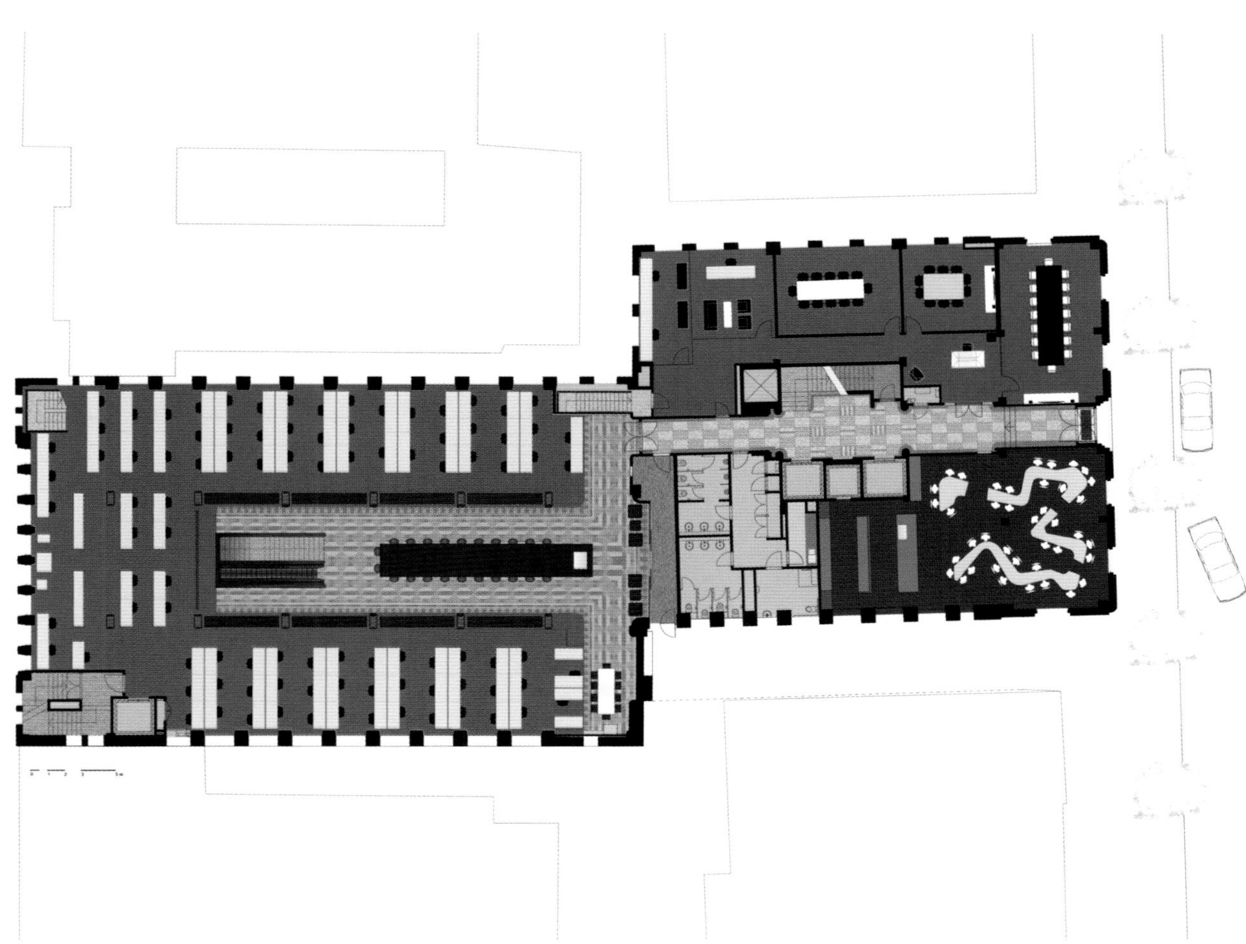

0 1 2 3 5m

0 1 2 3 5 m

Office Ogilvy

Design Company: Studio Ramin Visch

Project Location:Amsterdam

Area: 6300 m²

Photographer: Jeroen Musch

By definition an advertising agency that is proud of its creativity will not want to open up premises in any 'normal' neatly-raked business park. Nothing is more deadly to one´s image than a boring, commercially viable, common-or-garden rented office building. Disused warehouses, canal houses, churches, the catacombs of a football stadium and old factory premises offer a 'unique' artistic bohemian environment in which publicity people can flourish anywhere in the world. Ogilvy´s new premises in Amsterdam are no exception to this rule.

The initial idea for the reallocation of the former cycle factory was to split it up into commercially viable combinations of offices and manufacturing areas. Commissioned by property developer TCN Property Projects, the architectural firm of Neutelings Riedijk designed three new small office buildings to be completed this summer, to supplement the existing office wings of Simplex. On the advice of Witteveen and Visch, whom Ogilvy appointed as its architects/interior designers, the firm bought up all the factory areas of the complex in one go and had it upgraded to office space, thus giving the complex an entirely different set-up. Contrary to the original plan, the multi-occupancy building, now dubbed the 'A Factory', did not get any showrooms or stock rooms at all, but was exclusively reserved for office-type functions. To make the former factory halls of the cycle factory suitable for housing a publicity firm, architect Georg Witteveen and interior designer Ramin Visch devised a strategy of 'light urbanism'. In other words, there was to be no fixed allocation of the interior space with a permanent infrastructure of corridors and services; instead a more or less provisional exploration of the floor area was adopted, using detachable elements.

The concept of Witteveen and Visch was not based on any false romanticism about an authentic factory building, but on the idea of an adequate, 'open' working environment for the entirely non-hierarchical organizational structure of a huge advertising agency. The users themselves are responsible for the degree of cult in the interior. In this way many of the staff of Ogilvy cover the great distances in the former cycle factory, by means of Push-peddlers parked next to their worktables. It´s up to you to be creative, after all.

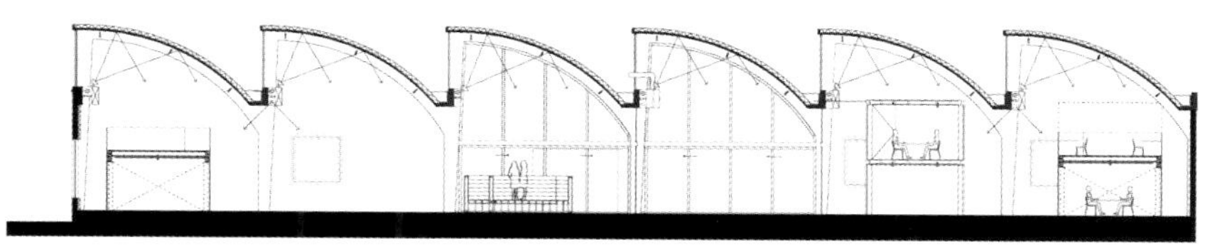

doorsnede AA

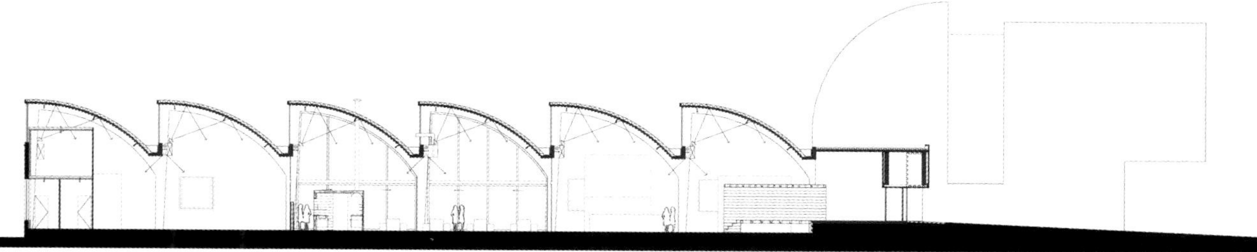

doorsnede BB

SCI-

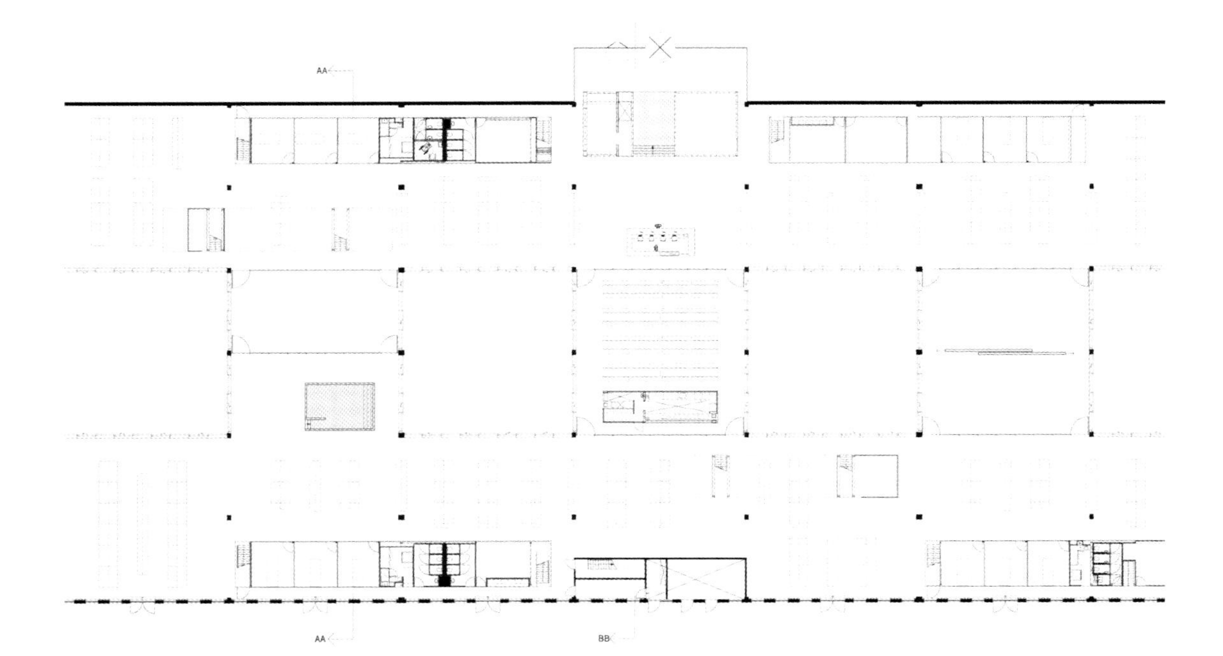
AA
AA
BB

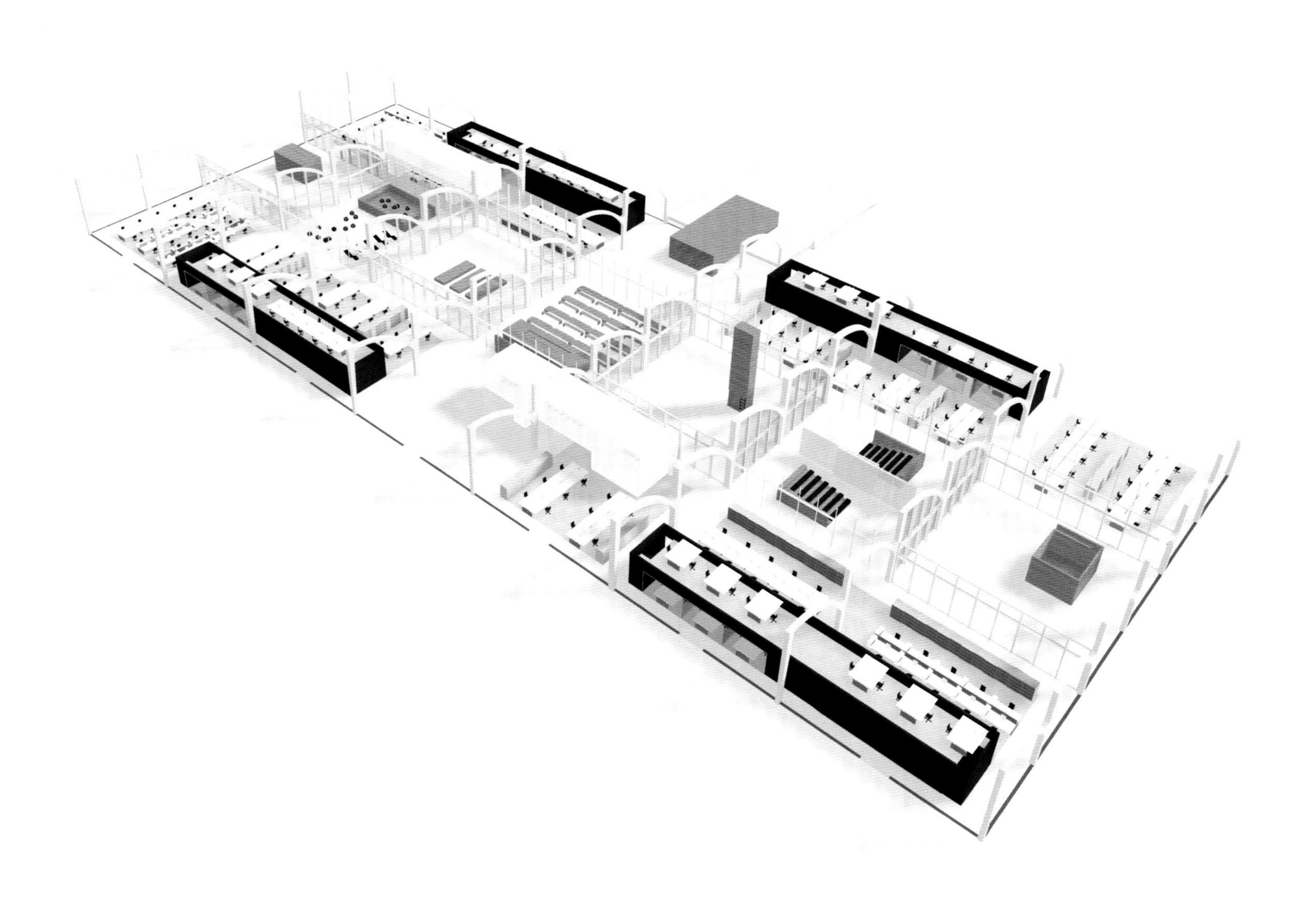

Office Besturenraad BKO

47

Design Company: COEN!

Project Location: Woerden, The Netherlands

Area : 2200 m^2

Photographer: COEN! | Roy van de Meulengraaf

COEN! created a new working environment and identity layer for the 'Besturenraad / BKO'. These two organizations are going to cooperate more intensively at a new location and take care of two denominational types of education in the Netherlands: Catholic and Protestant. The aim of this project was to visually connect the shared goals and principles of both organizations.

For the design of this story COEN! used The Book as a metaphor. Apart from the Christian and Catholic values a book also consists of structure, text and image. You see stained glass patterns, metal grids based on the golden section and special text prints with a message. The relation between faith and education is also subtly made clear by DNA patterns and golden 'office altars'.

Faithless

The eccentric and ornate atmosphere of religion has always fascinated a great many people. In addition, a new religion has developed with a belief in the global economy and with materialism as the one God. As people continue to search for new forms of spiritual enrichment, space is also created for new types of religious experience, Catholicism and spirituality. This search for new meaning leads to inner reflection. Coen van Ham delved into the tension that exists between the material and the spiritual. The result was a new logo based on the cross.

This logo is based on a reversal. Will you choose the deeper spiritual value or the material value on the surface? Do you believe in God, in yourself, or in the power of the economy? Or perhaps you believe in a material as well as a spiritual world?

Designer

Coen van Ham (1971) is a Dutch conceptual designer, architectural designer and source of creative inspiration. He studied at the renowned Design Academy in Eindhoven.

COEN! is one of the leading agencies in the Netherlands on account of its innovative concepts, stunning designs and inspiring workshops. His style is colourful, graphic, sober and communicating.

Coen and his team shape and guard over identity. Designs for your works of art, corporate identity, product and interior are all based in a powerful and unique concept. Designs by Coen van Ham have been included in several design and art collections. His work was shown in a diverse range of exhibitions worldwide. Coen's work has been featured in numerous magazines and books.

Hu: 7.0 If: 7.7 Kcv: 6.7 Kp: 8.8
Rek: 6.4 Sc: 7.2 Sk: 5.9 Te: 7.1
Cb: 6.0 Ckv: 3.2 Cumi: 9.5 Dr: 7.2
Hu: 6.3 If: 6.8 Kcv: 8.1 Kp: 6.0
Rek: 6.9 Sc: 5.4 Sk: 7.2 Te: 8.3

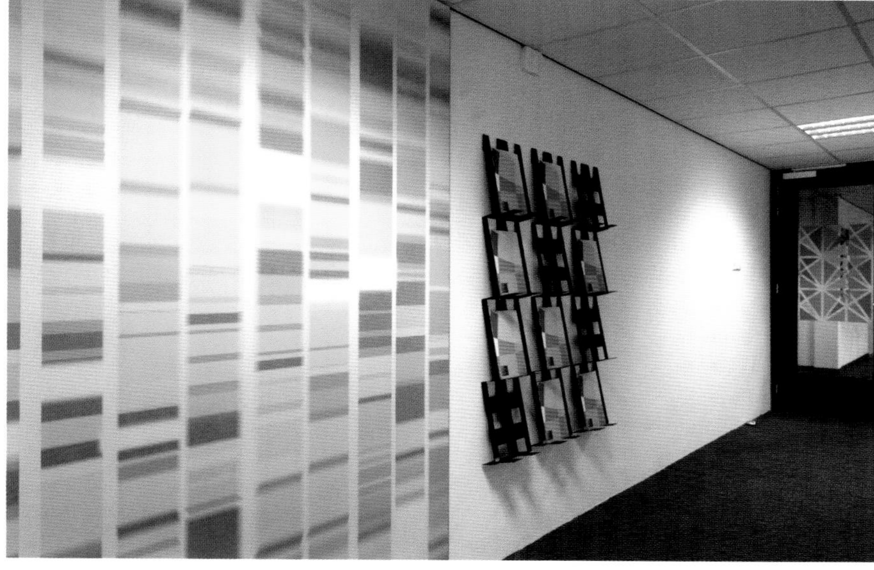

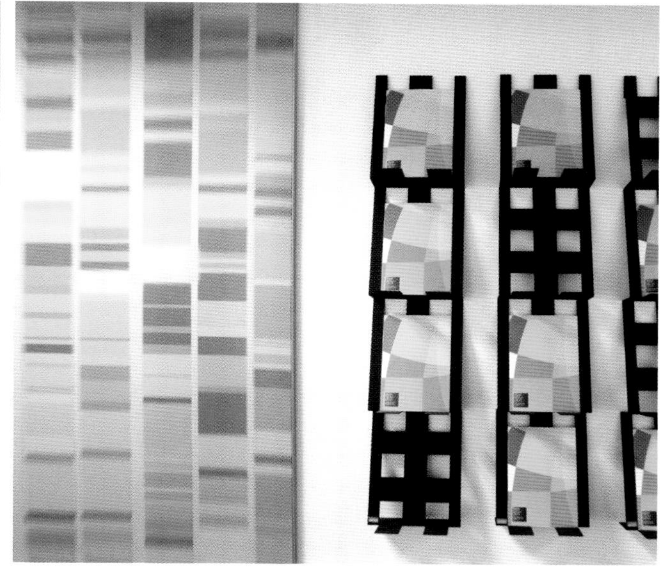

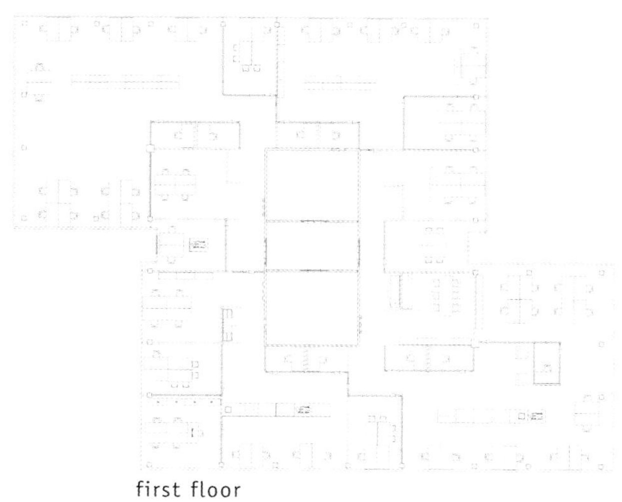

first floor

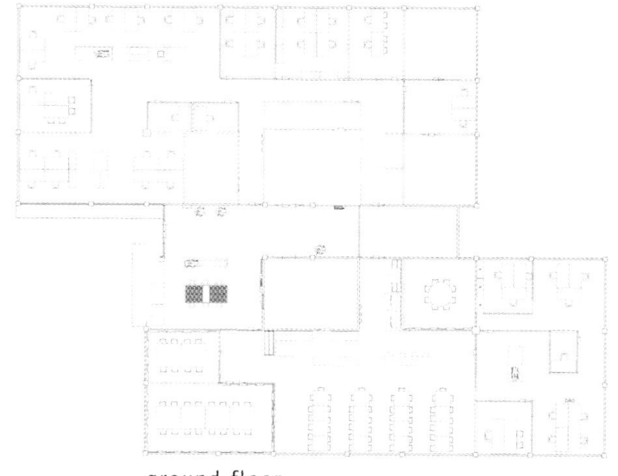

ground floor

Besturenraad | BKO

project	Besturenraad / BKO, Woerden, The Netherlands
design	COEN! bureau voor vormgeving
version	2011, copyright COEN!
scale	1:100

dagenzaal

Agency PUBLICMOTOR

48

Design Company: Bottega + Ehrhardt Architekten GmbH

Project Location:Stuttgart

Area :about 700 m^2

Photographer: David Franck Photographie

A loftspace with an overall size of about 700 m^2 in the center of Stuttgart was transformed into the new office space for PUBLICMOTOR.
One enters the office by passing through a `forest` of triangular, dark grey coloured volumes, leading the client to an open space and the waiting area, defined by a golden circle on the white floor. Small LCDs are integrated in the backside of the triangular volumes, showing projects and works of the agency. Rectangular volumes, covered with a light grey carpet, divide the space into seperated workstations. The volumes serve as sideboards, as sound insulation and have integrated lighting. The workstations of the management and the meeting room are divided by floor to ceiling - glass partitions and can be individually modified with bronze coloured, semitransparent curtains.
Yellow, vertical volumes, serving as the offices library, seperate the working area from the communication zone. A long wooden table, where the office comes together for lunch, is connected to a small kitchen in the back of the space. Behind a carpet covered wall, the service zone contains the cutting rooms of the agency and the restrooms.
A homogeneous white polyurethan floor emphasizes the openess of the overall space of the new agency.

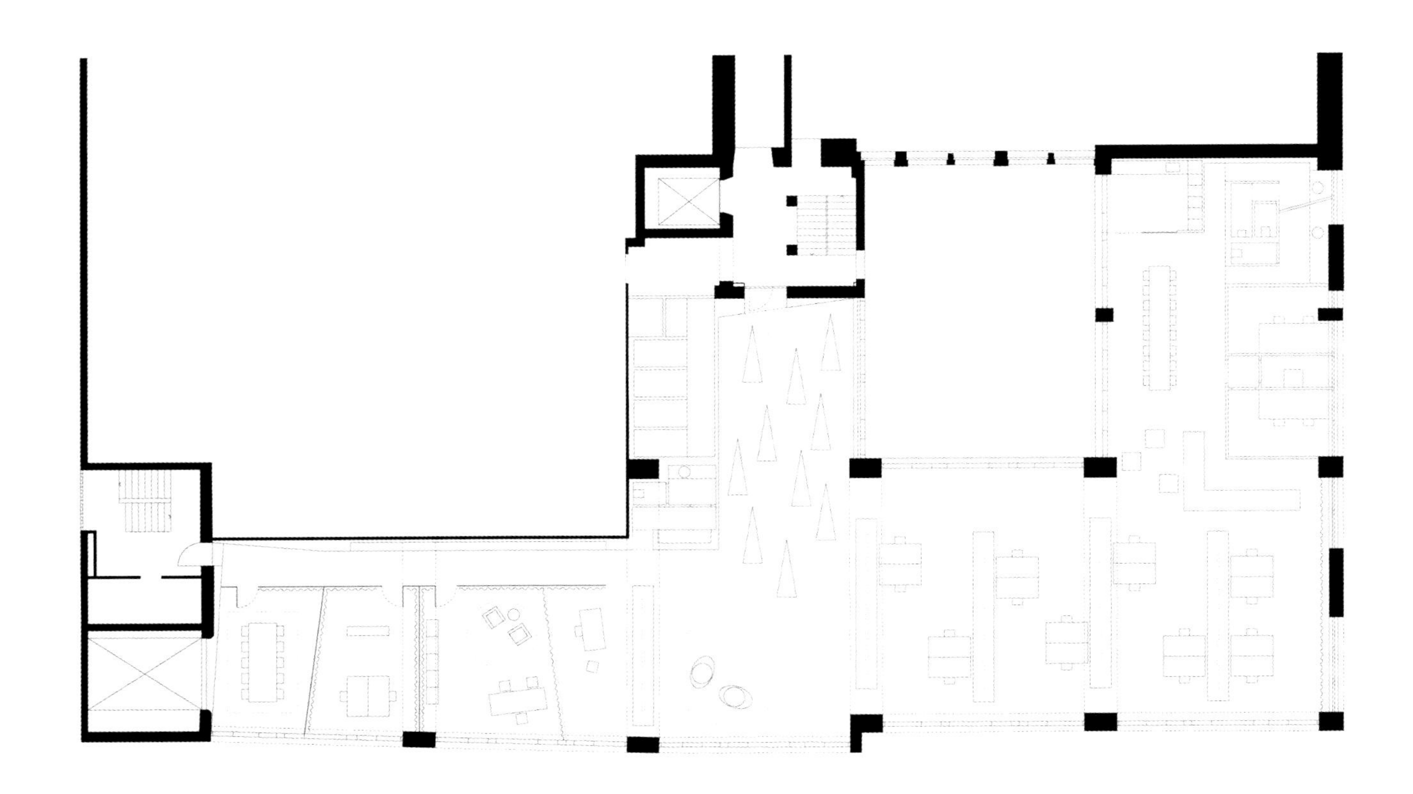

Jacques'

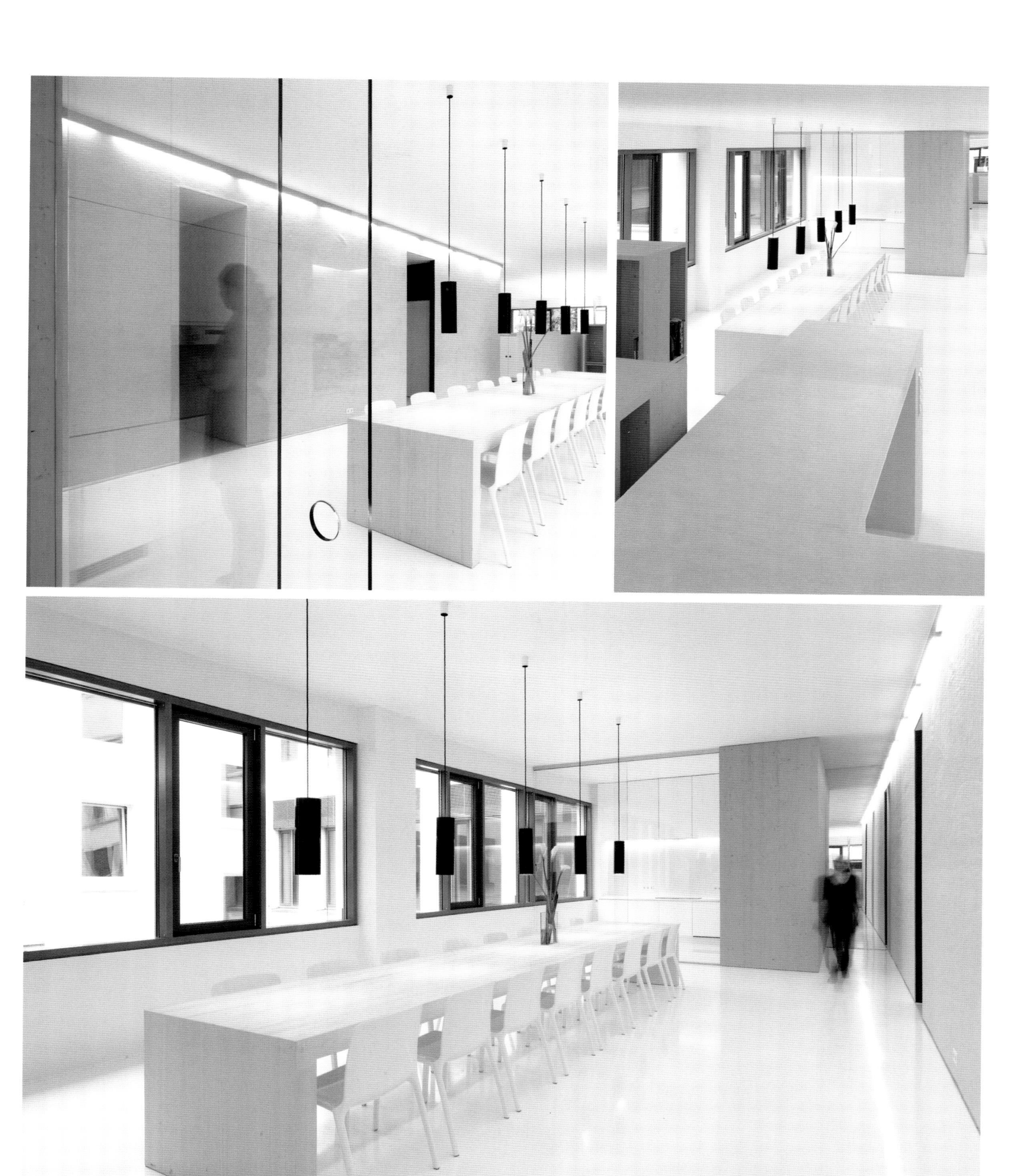

TOYOTA GAZOO.COM VIP ROOM

Design Company: PROPELLER DESIGN INC

Project Location:NAGOYA,JAPAN

Area :1020.5 m^2

Photographer: Nacása &Partners Inc.

TOYOA GAZOO.COM" is a new division developed by Toyota Motor and engaged in the "e-commerce business" in general. Toyota Motor Corporation, one of the world's top business enterprises, is launching a full-scale internet and trance-industrial product-distribution information business in the cybernetics market.

The products to be handled by the company include not only automobiles but every kind of product, along with the function of music distribution through internet. The company is located at 42F in JR Nagoya Station Building "JR Central Towers", a recently constructed new landmark in Nagoya. Its office commands an entire view of Toyota Automobile Museum standing at the place where Toyota Motor Corporation first started its business.

The office is more than 1300m^2 wide and will further be expanded in future.

The internal office space contains web contains, e-oft development room, kiosk terminal development / operation monitoring, etc. and transmits information toward the world.

The design concept used in the office construction can be shown as "e=(evolution of [e-commerce])-office", and intends to position the office space as the space media to create and transmit to the world the future forms, enterprise images and know-how E-commerce services offered by Toyota Gazoo for integrating the "Rigorism" and "Style" (i.e.,dignity) as an enterprise and the "Flexibility" and "Casualism" (i.e.,casualism) as a venture business.

The design method used expresses this concept of integration as "Material Complex + α" ,i.e., grids formed with metal and wood with different tastes piled up one on top of another(the Material Complex) plus fresh (+ α) produced, together with delicate lights through indirect illumination, by the glass gates incorporated into the wood walls wrapping up internal space like hum an skins.

GAZOO.com GAZOO.com GAZOO.com

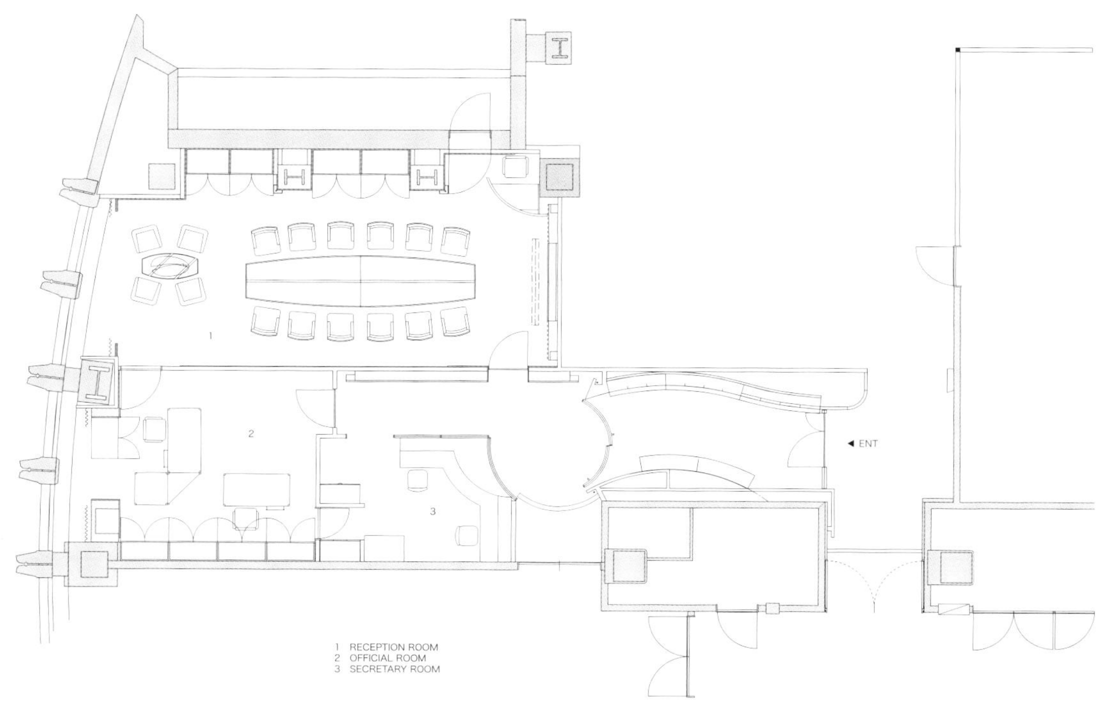

1 RECEPTION ROOM
2 OFFICIAL ROOM
3 SECRETARY ROOM

GAZOO.com GAZOO.com GAZOO

TOYOTAGAZOO.com

Paga Todo

Design Company: USOarquitectura

Project Location:Mexico

Area: 2000 m²

Photographer: Héctor Armando Herrera

Space is the main factor that determines the interior design and operation of a company. The new corporate offices for Paga Todo presented a particular challenge because it was necessary to adapt to the clients demands and a 2,000 sq m area in a shopping center.

A big wood box, inserted respecting the surrounding design, greets everyday collaborators and visitors. Inside the box were located the reception, support area and interview halls, on top of it –with a panoramic view of the finance area- the personalized area to serve the dealers.

The client decided to implement in their office a lounge style cafeteria –like a hotel lobby- because before the relocation the majority of the collaborators preferred to work and meet in the close by cafeterias to enjoy a more relaxed ambiance. This space has all the necessary services and it is a nice surprise for the visitors because there are screen, complimentary computers with Internet, snacks and drinks.

The staircase was located in the vertex of the project in order to communicate with the upper level, opening a new entrance of light from above and making more interesting this meeting point for the colleagues.

The color palette –asked by the client- is very sober and with no risk. White, beige shades with accents in a dry green and the oak of the furniture and woodwork. Three sections with meeting halls divide the space generating references and transitions between the work cells.

For natural light big vertical stripes were open on the façade of the shopping center and most of the walls were not built ceiling height to make the most of the different natural light sources of the building. Large windows facing the interior of the shopping center were also installed to make references in the main corridor. The windows have random size and create a sequence with the transition of each of the work teams.

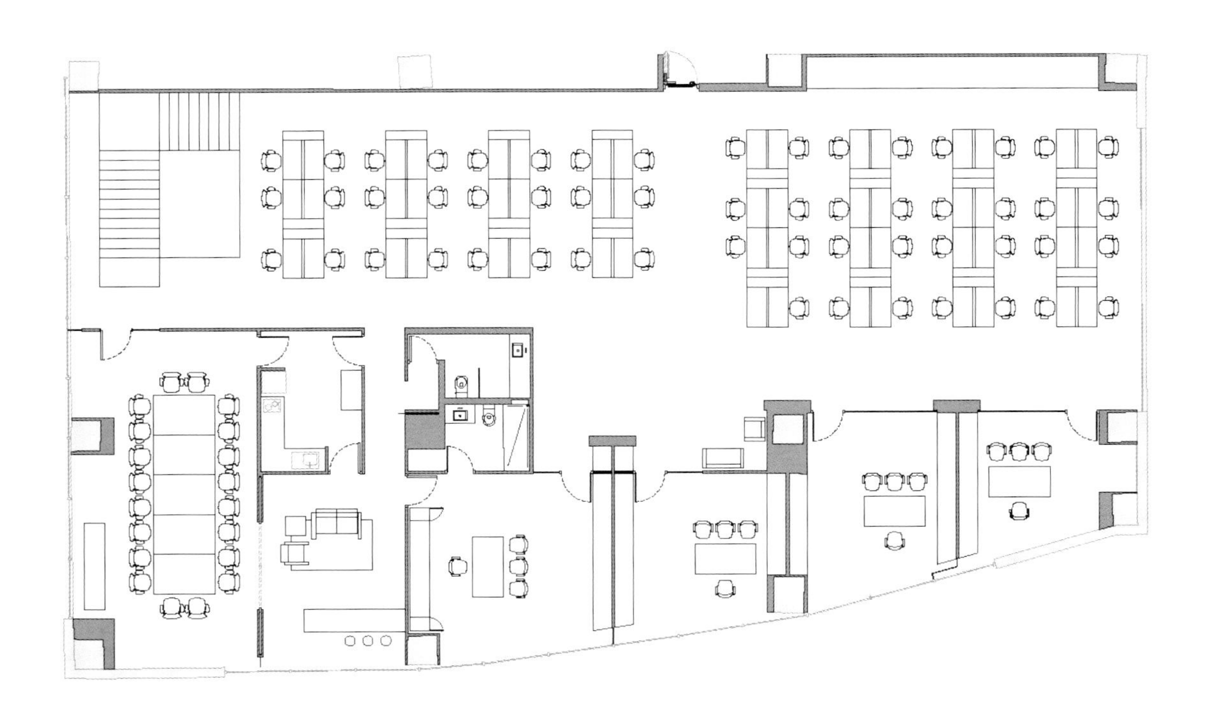